HOLLYWOOD'S LOST BACKLOT

HOLLYWOOD'S LOST BACKLOT

40 Acres of Glamour and Mystery

STEVEN BINGEN
WITH MARC WANAMAKER, BISON ARCHIVES
FOREWORD BY RON HOWARD

Guilford, Connecticut

An imprint of The Rowman & Littlefield Publishing Group, Inc.
4501 Forbes Blvd., Ste. 200
Lanham, MD 20706
www.rowman.com

Distributed by NATIONAL BOOK NETWORK

Copyright © 2019 by Steven Bingen

All photographs and maps courtesy of Marc Wanamaker, Bison Archives, unless otherwise credited.

All rights reserved. No part of this book may be reproduced in any form or by any electronic or mechanical means, including information storage and retrieval systems, without written permission from the publisher, except by a reviewer who may quote passages in a review.

British Library Cataloguing in Publication Information available

Library of Congress Cataloging-in-Publication Data available

ISBN 978-1-4930-3361-4 (paperback)
ISBN 978-1-4930-3362-1 (e-book)

∞™ The paper used in this publication meets the minimum requirements of American National Standard for Information Sciences—Permanence of Paper for Printed Library Materials, ANSI/NISO Z39.48-1992.

Printed in the United States of America

Contents

FOREWORD BY RON HOWARD	ix
PREFACE: LOST ACRES	xi
CHAPTER 1: THE MAN WITH THE MEGAPHONE	1
CHAPTER 2: MOGULS, MILLIONAIRES . . . AND A BANDLEADER	15
Cecil B. DeMille	15
RKO	20
David O. Selznick	24
Desilu	35
CHAPTER 3: FRONT LOT: SETS, SETTINGS, AND SET PIECES	40
1. The Mansion	43
2. Stage 1	47
3. Postproduction	49
Stages 2, 3, and 4	
4. Stage 2	53
5. Stage 3	57
6. Stage 4	62

7. Bungalows and Offices	64
8. Mill	69
9. Wardrobe	70
10. Commissary	72
11. Property	74
12. Mail Room, First Aid, and Security (later Stages 5 and 6)	77

Stages 7, 8, and 9

13. Stage 7	80
14. Stage 8	82
15. Stage 9	85

Stages 11, 12, and 14

16. Stage 11	88
17. Stage 12	93
18. Stage 14	96

Stages 15 and 16

19. Stage 15	100
20. Stage 16	104
21. Makeup and Dressing Rooms	107
22. Water Tower	109
23. Stage 10	110
24. Scene Dock	112

CHAPTER 4: **40 ACRES: HOLLYWOOD IS A FACADE** — 115

1. Ince Gate	115
2. Farm and Barn (later Camp Henderson)	118
3. Tarzan Jungle	122
4. Arab Village and Medieval Village	125
5. Western Street	132
6. Tara (later Jerusalem, Stalag 13, and *The Fortune* sets)	136
7. Railroad Station Set (originally the Temple of Jerusalem)	155
8. Country Home	168
9. Location Roads	170

10. City Streets	173
11. Barracks and Dock Sets	191
12. Residential Street	194
13. Reform School Set and Bombed Town (and later Alley)	197

CHAPTER 5: THE MEDIA CAMPUS — 203

NOTES	225
BIBLIOGRAPHY	228
INDEX	234
ABOUT THE AUTHOR	247

Foreword

I REMEMBER 40 Acres.

I must have been about six years old when I first visited Mayberry, or rather 40 Acres, for *The Andy Griffith Show*.

The town of Mayberry, North Carolina, had actually been introduced earlier on an episode of *Make Room for Daddy*, which I haven't seen in years, and barely even remember. But I remember the *Griffith* show. I remember 40 Acres.

Television was made very quickly then. We used to make an entire episode of *Andy Griffith* in three *days*! The 40 Acres scenes were usually shot on Mondays. For some episodes we would get all of the 40 Acres stuff in just half a day. Ultimately, we did 249 episodes of *The Andy Griffith Show*, most of them utilizing 40 Acres.

The interiors were shot on Stages 1 and 2 at the Desilu Cahuenga studios in Hollywood—which are still there. The backboard, if not the hoop I used to shoot basketballs into on the stage, is still there too, nailed to a wall—at least it was the last time I looked for it. *I Love Lucy*, *The Dick Van Dyke Show*, *Make Room for Daddy*, *The Real McCoys*, and *Hogan's Heroes* were all on the lot when we were, or before. More recently *The Artist* used the lot, which is now called Red Studios. You can still visit the place.

40 Acres, however, now exists only on-screen, and as I remember it. That's probably why I still find myself remembering it so vividly, and so often. In fact, my wife sometimes gets mad at me because, even after all these years, I keep remembering, keep talking about that place. About 40 Acres.

40 Acres was a movie backlot, but for me it was a playground as well. I'd ride around the lot on these paved and unpaved roads that circled the property on my

prop bicycle, the same one Opie drove on the show. My parents wouldn't let me cross Ballona Creek, so I never knew that an artificial lake, and even Tarzan's Jungle, stood over there. I sure would have liked to have seen them. But everything else, every place else on the lot was there seemingly just for me to explore.

I was a fan of *The Real McCoys*, so I remember getting a kick out of seeing and recognizing that farmhouse. But sometimes I was deceived, or misremembered, what shot where. For years I assumed that the *Hogan's Heroes* set I liked so much had been built for Billy Wilder's *Stalog 17*—it wasn't. George Stevens shot *The Greatest Story Ever Told* on the lot when we were there, and I recall exploring those enormous sets as well. Yet I always remembered that they were instead built for DeMille's *Ten Commandments*. I didn't know until now that some of those sets did, in fact, go all the way back to DeMille, if not quite to actual Biblical times.

All of this is what backlots are created to do. They make us believe they are something they are not. I appreciate backlots very much. I did *The Music Man*, and later *The Shootist*, on the same Midwest Street at Warner Bros. One of my first films, *Deed of Daring Do*, was shot on the Western Street at CBS. I won $100 dollars in a Kodak-sponsored contest for that one, so it could be said that my career as a filmmaker began on a backlot.

Even today, when I get the chance, I like to explore the surviving studio backlots. I enjoy their eccentric energy and their immense, improbable combination of both ersatz and actual history—history written on 2x4's and on wooden flats, history exposed on film, and, finally, upon memory.

Yup, I remember 40 Acres. I'm quite sure I'll always remember 40 Acres. And I'm very gratified that some of you, now, will as well.

—Ron Howard
July 2018

Preface: Lost Acres

Last night I dreamt I went to Manderley again. It seemed to me I stood by the iron gate leading to the drive, and for a while I could not enter, for the way was barred to me. Then, like all dreams, I was possessed of a sudden with supernatural powers and passed like a spirit through the barrier before me . . .

—THE SECOND MRS. DE WINTER, *REBECCA*

LET'S GET SOMETHING STRAIGHT right away. It was not, in fact, in "Hollywood" at all.

Oh, and its size. The backlot's size was not really 40 acres—despite its always being called that.

And in reality, the place was not, technically, a backlot at all. Because it was not, technically, attached to the back of a studio property.

The place is, however, truly and irretrievably and forever lost. That part of our title, sadly, cannot be argued with.

One more thing. Unlike my other studio volumes, on MGM, Warner Bros., and Paramount, the facility under discussion here, actually located in Culver City, California, was not really a traditional movie studio at all.

Make no mistake, it was used for movie production. Some of the most famous movies and television productions of all time were shot behind its fences. But the names painted on those fences changed many times over the plant's convoluted, and largely unchronicled, nearly sixty-year lifespan.

Unlike the previously mentioned companies, the 40 Acres backlot (and let's just agree, irony-free from now on, to call it that) was a film factory in the truest sense of the word. The above companies, and the other Hollywood studios, used, and continue to use, their plants as campuses and as corporate headquarters, as well as factories. In fact, in twenty-first-century Hollywood, the "studios" are corporate sites first, symbols more than soundstages, boardrooms more than backlots, and are used as production facilities often only when there is nowhere else cheaper to shoot available.

40 Acres, however, or at least the backlot sector, which was, as mentioned, separate from the rest of the plant, was just a place where movies were made. There were no offices or departments or recording facilities or barbershops or commissaries or gymnasiums. There were not even any permanent soundstages. It was *only* a backlot. It was only a place where things were *filmed.* The backlot iconography on the site was undiluted and unmistakable: fake streets, flats held up by scaffolding, towns without people. Western villages next to Roman temples and South Seas jungles. That, and that alone, was its raison d'être.

Unlike other studios, 40 Acres, front *and* backlot, had no overriding, decades-long affiliations. Its only permanent identity was that of the movies—and later of the television shows—ground out like yellow cornmeal on the site. Thomas Ince, Cecil B. DeMille, Joseph P. Kennedy, David O. Selznick, Howard Hughes, and even Lucille Ball all controlled the property, but none of them managed to bend it to their wills, to make it theirs for very long. Instead, it was the other way around. The studio, in fact, shaped their actions and their projects. It made them all its patrons.

It's all gone now. The other studios exist, at least as corporate identities. But 40 Acres—let me emphasize again that now I am referring to the backlot property only, even though the entire lot is going to be covered in the text—is long gone, both as a physical place and as a corporate one. RKO, the longest-lasting owner of the property, is the only major Hollywood studio ever to cease active production and basically just disappear, although the name will undoubtedly continue to exist as a legal bookmark for the rest of time. Likewise, the former 40 Acres itself, where a thousand movies were made, is now an industrial site where, with no apparent irony, TV shows are still being produced.

The trail to follow in telling the 40 Acres story has, to put it optimistically, long gone cold.

Fortunately, when I started down that trail I had a lot of friends to light the way. 40 Acres has developed a fervent cult following since its demise, which is truly remarkable. Eventually, certainly unknowingly, perhaps derisively at first, I moved from studying this cult to sharing in its odd obsessions. 40 Acres is a place I have never walked. Yet, like the second Mrs. de Winter, I've now passed ghostlike through its gates hundreds of times in my dreams. In these dreams I've explored its nostalgic small towns and film noir ghettos and exotic foreign villages and dark jungles. I've made dubious deductions about its many mysteries. I've romanticized the sometimes-cynical cavaliers who created it, and raged at the barbarians who destroyed it. Part of me, I fear, will now never escape 40 Acres. And it worries me just a bit that I'm okay with that.

Fortunately, the other people inside this strangest of cults have been, without exception, willing to let me join them not only beyond those spectral gates, but even in their actual homes. They have generously shared information and insights, and artifacts and anecdotes, and have thrown open the doors to their collections and let me take the conclusions and connections that came out of their hard work and pass them off as my own.

So, as tempting as it is to take advantage of all of this generosity, I feel compelled, in telling the story of the place that gave the world the likes of the honorable Ashley Wilkes, Superman, Andy Taylor, and Spock, to do the honorable thing and share the names of the following:

John Bertram was my initial entrée through the gates of 40 Acres, having worked at Culver Studios himself, having done a lot of research on the property, and having then generously shared that research with me. Both John and Margi Bertram, in fact, spent countless hours meticulously studying old production paperwork, initially on behalf of the studio itself, which they later kindly let me pillage. Many of the previously unearthed details printed here for the first time came from their exhaustive efforts.

No studio exists in a vacuum. Hollywood is an incestuous place, with one studio's tentacles clutching and coiling and ransacking the vaults of another. For example, Walt Disney archivist emeritus David R. Smith generously set me straight as to Culver Studios' contribution to Disney's *Fantasia*. Likewise, Colin Greene, Sony Pictures VP of Production Asset Management, marveled with me over the strange possible fate of *Gone with the Wind*'s Tara. Speaking of which, Peter Bonner, the colorful Georgia curator who holds the destiny of Tara in his capable hands, and Tommy Jones, who has well-chronicled that destiny, were also both very helpful during my own admittedly quixotic quest for the most famous "fictional" house in the history of the world.

Rex McGee explored the backlot himself in its twilight and generously provided me with high-resolution photographs of what he saw. I'm happy to be able to print those photographs here for the first time ever. Thank you, Rex.

Eminent historian and illustrator Joseph Musso, who was lucky enough to explore 40 Acres when it was still intact, during his many years in the film industry, answered a thousand questions about the property for me and shared with me the story of Walter O'Connor. O'Connor, it seems, once happily worked at the studio as a historian and long-dreamed of telling its amazing story in print someday himself. O'Connor tragically died from an auto accident the same week I contacted Joe for my own book. I can only say that I hope Mr. O'Connor's inadvertent, and sadly postmortem, contributions to my book will do him proud. I really do.

Joe also introduced me to art director William J. Creber, who shared some unforgettable and never-told stories about *The Greatest Story Ever Told*.

Charlie Ziarko kindly shared stories with me about the studio over dinner one night at Hollywood's legendary Musso & Frank Grill. Those stories, and his ultimate input into the form and flow of the final book, cannot be overestimated. Next time, Charlie, I might even pick up the check.

Kipp Teague and Jake Shepherd have both done groundbreaking work on the 40 Acres story. Kipp's website and Jake's book, both gratefully referenced in the bibliography, are the points of light from which any future work on the subject can begin to be illuminated. Their efforts, and their personal support and involvement, shaped this volume in countless ways.

Robert Lane was responsible for the architectural aspects of the book, meaning that it was he who was behind the crafting, drafting, or adapting of many of the maps seen within its pages. Rob has done this for me in all of my studio volumes, which have all benefited immeasurably from his efforts.

Grateful appreciation is also owed to Louisa Velis at Imagine Entertainment, to historian/sound designer Ben Burtt, and to director Ron Howard, for remembering 40 Acres.

A book of this nature could never be completed without the assistance of archives and special collections around the world. Michael L. Gilmore, of the Harry Ransom Center at the University of Texas at Austin, was invaluable in helping me wade through the David O. Selznick archives and unlock a few of its many secrets. Louise Hilton, of the Academy of Motion Picture Arts and Sciences Margaret Herrick Library, again found what I wanted, and what I needed, even before I knew I needed it, and had it placed on my desk there. Thank you again, Louise. The staff of the UCLA Special Collections Library, specifically Amy S. Wong, were also very helpful in assisting me with the RKO production records archived there, as was Brett Service, curator of the USC Warner Bros. Archive.

Speaking of archives, Marc Wanamaker is the ultimate contributor to have in any Hollywood work. I'm so grateful that he has repeatedly, over several years now, trusted me with the evaluation and preservation of the countless priceless artifacts from within the vaults of Bison Archives, which he has generously shared with me again and again.

Additionally, I'd like to acknowledge contributions by the following individuals who were important enough to be mentioned here by name, if not, perhaps, important enough for me to remember why. Regardless, the following know what they did to contribute to this book, and I'm grateful, so much, for all of their important, if unspecific, contributions. This paragraph is for you, Marilyn Allen, Ron Barblagallo, Craig Barron, Steve Bauer, John Bengston, Billy Blackburn, Mischa

Hof, Rob Klein, Nancy Knechtel, Tom Loveman, Mary Mallory, Mark A. Martin, Jarod Millsap, Ray Morton, E. J. Stephens, Steven X. Sylvester, Stan Taffel, Frank Thompson, and David Warden.

I'd also like to mention my parents. And Beth Orsoff and Zoe Bingen. They know why.

So, my friends, I now bid you to join me behind those gates. I very much hope your journey is as interesting and eventful as mine has been. As you explore, though, I'd like you to note, once or twice, how, unlike any other historic journey, be it into past worlds or lands or lives, the 40 Acres experience is truly, quietly unique. Because, of course, like the movies made there, nothing behind those gates was *ever* real.

—Steven Bingen
Hollywood, California, 2018

CHAPTER 1

The Man with the Megaphone

Filmmaking is the process of turning money into light, and back into money again.

—John Boorman

THE STORY of human civilization in America starts with the arrival of the Native Americans. Not to make comparisons, but so too does our story.

The Indians in our story were real Indians, maybe—as Oglala extras were, in fact, available to a particular producer filming a particular movie here more than a hundred years ago. If this was the case, the Indians would have brought their own costumes, although these same Indians were apparently not above raiding the wardrobe and prop bins back at the studio in search of colorful pieces with which to add a dubious authenticity to their buckskins.

However, these Indians may actually have not been Indians at all, but rather Caucasian, or Mexican, or even Asian actors that day—awkwardly rowing birch-bark canoes up Ballona Creek—in which case these ersatz Native Americans would have been wearing dime-novel costumes from those same bins and itchy, braided wigs.

In any case, the creek itself, if nothing else, certainly was authentic. It turns out that that creek had been selected because the Los Angeles River, and specifically the terrain around it, was unsuitable and because the Pacific Ocean, regrettably, was the only body of water fronting the studio that the company had driven out from.

The year was 1915, and a movie, the first of thousands to be birthed on that spot, was being made. A hundred-plus years later, no one seems to even remember

Thomas Ince, flanked by cowboys and Indians, including leading man William S. Hart, who stands to his left. All look to the future. (1914)

what the title of the film was. The important thing to remember here is rather that those possibly apocryphal Indians floundering across those waters that day were observed by a very interested spectator.

His name was Harry Hazel Culver. And he liked what he saw.

Culver was a real estate developer from Nebraska. He had appeared in California in 1910 with ambitions of creating a community with his name on it, and at this point he was well on his way to making those ambitions happen. Although "Culver City," as it would come to be known, with its original population of 550, would not be incorporated for another two years.

Culver was more interested in the men on the bank filming the tableau than what was actually being filmed. He had noted with a most keen interest recently the proliferation of movie crews that had been marauding across Southern California in recent years. Most respectable and conservative businessmen of Culver's standing were disdainful of the movies and the people who watched them, and most certainly of the people who made them. But Culver, with his middle-class, Midwestern sensibility, very much liked going to the photoplays, liked seeing them with audiences of laughing immigrants, many of whom, he was smart enough to realize, would eventually be joining the middle class themselves and hopefully, soon after, buying his houses.

He also liked, within reason, the people who made those movies. Many of them were themselves immigrants, or the children of immigrants: Eastern Europeans, possibly even Jewish. And make no mistake, they were loud, loud and boastful, and crass and profane—East Coast types of people to be sure. Yet one had to hand it to them: They were inventing an *entirely new industry*. And no one in Los Angeles, except him, seemed to bother to notice this.

Many of these picture people were already clustered in a nearby village named Hollywood, yet the residents there, strict prohibitionists, treated them like they all suffered from smallpox. Hollywood had even gone so far as to scornfully pass laws making it illegal to show movies publicly within that community.

Harry Culver walked over and struck up a conversation with the man in charge of the little crew. He found that man easily enough. He was the fellow who yelled the loudest—and through a megaphone yet. Culver convinced that man that his community, when created, would welcome—indeed would provide completely free, or at least on a deferred payment plan—real estate to any producer willing to relocate to the Ballona Creek area.

That man in charge, the man with the megaphone, was one Thomas Harper Ince.

Ince was a man whose importance to the early development of the motion picture *industry* cannot be overstated. Born in 1880 of English stock in Rhode Island, to a show business family which Culver might not have entirely approved of, Ince had drifted into the fledgling film industry in 1910—initially because he had a young family to support, and because an old theater colleague, one Joseph Smiley, was willing to recommended him for a job.

His early days in the movies were spent as an actor, but Ince, among the first of millions to follow, quickly fell in love with a business he had originally seen only as a get-rich-quick scheme. He quickly realized that the real way to be creative in the movies was on the other side, the narrow-concentrated side, of that megaphone. "The story-teller has never found so pliant a medium for his interesting occupation as the screen," he would pompously pronounce in 1916. "Drama has long been a thing of the artificially lighted stage, and it has been so bound in by time, by the mechanical exigencies of scene and costume, by the limit of human adaptability and endurance, that it is little short of marvelous that such complete representations have been made."[1]

So Ince rapidly became a director, and then one of the first director-producers. What's more, he turned out to be a keenly creative producer with an eye for story, location, and visual composition, and for spotting talent.

He also created, not incidentally, the business model that movie studios, to a large part, still operate by today. Before Thomas Ince, those movie studios, if they could even be called that, were disorganized and hurly-burly and chaotic. There was no cohesive production process in place—so everyone, from the actors to the visiting landlord shuffling around looking for his rent, might be asked to pitch in, moving sets or carrying equipment, all just to try to get the damn thing done.

Compounding this chaos, there were usually no scripts, or scenarios as they would have been called then, to determine what the story was, and no one on the payroll to write it all down anyway. Usually, the on-screen action ended up being improvised on the set—also improvised—and the whole tottering, unwieldy process would lurch along, from inception to distribution, stumbling against all reasonable odds from one to the other, with no apparent plan in mind along the way.

So it was Thomas Ince who decided that there had to be another way. He based his ideas about film production not on the model of how films had chaotically and haphazardly been glued together in the past, but rather on that of the heavy industry of Detroit, where automobiles were constructed from first nut to final paint job in one place, with the same people doing the same job on each vehicle that ultimately rolled glistening out the gate.

Ince built his first studio, eventually called "Inceville," in 1911–12 at the end of Sunset Boulevard on the Pacific Coast Highway in what is now the Pacific Palisades. Filmmaking concepts that later became common at all the studios were first implemented in this foggy, windswept crevice above the Pacific Ocean.

However, in spite of the grandeur of its conception, Inceville in execution resembled nothing so much as "a sleepy, dirty western town—scattered buildings, of plain boards, and rut-worn roads leading into the hills,"[2] as actor John Gilbert unsentimentally remembered it. Yet it was here that the organized, departmentalized, and later unionized concepts that would turn studios into empires were created. It's also worth noting that it was here, for the first time, that standing sets were intentionally left standing, on a so-designated spot at the back of the studio grounds, effectively making Ince the father of the backlot—although father Ince once fibbed in an interview that he, in fact, seldom reused sets, believing, he said, that they would be recognized by audiences.

Ince is also credited with inventing the title of "production manager" and formalizing the functions of the editor, screenwriter, costumer, accountant, and even director (the director until this time had been forced to assume the responsibilities of all the above), which, as he had hoped, enabled his hardscrabble little studio to make more than one film at a time. A film industry first—in what was not yet an industry at all.

Inceville is where many of the concepts by which motion picture studios still operate were first implemented. Behind the clapboard bungalows and false fronts to the south is Santa Monica Bay. (1919)

Inceville didn't last long. Although he created stars and directors, and even the very concept of stars and directors, and produced hundreds of films on the site, Ince knew that the property, if not the production methods created there, was too distant from the rest of the city to ever be practical, and the dampness and gloom that washed in from the ocean was hard on both the sets and on the actors using them. What's more, the wind often blew that dampness and gloom as well as, even worse, sand into the laboratory equipment, making much of the precious film—the result of everyone's efforts—being processed there that day unusable.

In 1915 a fire nearly burned the studio down, and early the following year, a second, more extensive blaze destroyed some of the property as well. Ince, who at the time was finalizing a deal for a venture called the Triangle Film Corporation with fellow producers Mack Sennett and D. W. Griffith, was understandably considering a move, perhaps to Hollywood, when, megaphone in hand on the banks of Ballona Creek, he happened to meet Harry Culver.

Post-Ince, Inceville continued to be used as a studio by, among others, Ince discovery William S. Hart and Robertson-Cole Pictures until 1922. Two years later the remaining facades, which must have looked like an actual Western ghost town

THE MAN WITH THE MEGAPHONE 5

The chaotic-appearing "stages" at Inceville were open to accommodate available light, which was often in short supply on the hazy Pacific coast. (1913)

by this point, caught fire one last time, scattering their blackened dust into the ocean.

Ince and his partners built a new studio in the still unincorporated "Culver City" in an appropriately triangular shape fronting Washington Boulevard. Some of the property had already been used as an army barracks, but the impressive Corinthian-columned colonnade facing the street was probably Ince's idea. The producer wanted his next kingdom, unlike his last, to be impressive looking, at least from the street, as well as functional. In March of 1916 the studio was officially opened.

The colonnade is still standing, but the partnership and Triangle Pictures was not to be long-lasting. A major point of contention between Ince and the company was the studio itself, which Ince owned, as per his deal with Culver, but which management asserted belonged to the company and even listed it as a company asset in their stock offerings.

The new Triangle Studios in 1916. Note the prominence of Ince's name in the signage. The street running along the side of the lot on the left has since been swallowed up by studio expansion, but significantly enough, was named Ince Way.

In 1917 Ince sold his interests in Triangle and, reluctantly, in the Triangle Studio itself. The studio was purchased in 1918 by Goldwyn Pictures, which, through another siege of mergers and takeovers, eventually emerged as the mighty Metro-Goldwyn-Mayer, which it would remain until 1986. Today the property is Sony (Columbia) Studios. As noted, the colonnade is still there.

Harry Culver, standing on the sidelines during all of this blood sport, undoubtedly realized that Ince, down but certainly not yet out, would eventually need yet another property, this time to be called, finally, Thomas H. Ince Studios. More than happy to oblige, he offered him a 14-acre property that was an almost literal stone's throw from the Triangle lot.

That property had, since 1781, originally been part of El Pueblo de Nuestra Señora la Reina de los Ángeles, but it wasn't until 1821 that Rancho Rincón de los Bueyes was created on the grounds. The name translates as "corner for the oxen," because of a box canyon in which livestock could be herded. The Higuera family, which included the son of a former *alcalde* (mayor), claimed most of the land. In 1872 a United States patent listed the property as belonging to Francisco and Secundino Higuera and consisting of 3,100 acres.

According to historian Jacob N. Shepherd III, Achille Casserini, a Swiss immigrant, acquired at least some of the property, probably around the turn of the century, which would become the 40 Acres backlot, although Culver, at this time, was only interested in getting Ince to sign on the dotted line for the primary front lot parcel.

According to the document above that dotted line, which was recorded at the Los Angeles County Hall of Records on November 4, 1918, Culver agreed to build, and to pay for, the construction of the first three buildings on the property, namely what would soon be known as the Mansion and the first two stages, as well as "such other buildings" as needed, to the sum of $132,000. In return, Ince would lease that

An early architectural concept drawing for Ince's third studio. Except for the cameraman cranking away on the left the property could be a Virginia equestrian site. (1918)

property for five years for $90,000. He was also given the option to purchase the lot outright anytime within the first three years for $145,000, an amount that any rents paid to Culver to that point would be applied against.

The press, however, probably being fed inflated information by Ince's publicists, got most of the details wrong. For example, *Moving Picture World* reported in December 1917 that the price of the property was $25,000 and that "the plant to be erected on the site will cost approximately $300,000 and embrace eighteen buildings, built in Spanish mission style with an imposing facade on Washington Boulevard."[3] In June 1918 Culver formally announced that Ince would spend a somewhat scaled-down but still impressive $200,000 on the property. And yet a month later, on July 19, 1918, when it was reported that Ince had formally acquired the property on an "option basis," the $132,000 was listed as a "loan" from Culver, which, in this case, was rather accurate.

Immediately Ince made it clear that this studio, whatever its method of conception, was to be something special. He kept, somewhat surprisingly, the idea of that "imposing facade" on Washington Boulevard. And he, and his architects, spent an unusual amount of time planning the aesthetics of the studio, and the aesthetics that could be achieved by filming at that studio. For example, in December of that year, Ince's publicists announced to the trades that "the studios are situated on the main boulevard running from Los Angeles to Venice of America. So on the one side is the big city, on the other the ocean and to the rear is an imposing range of mountains. Picture atmosphere of every quality is conveniently available."[4]

In his supervision of the design of his new plant, Ince obviously tried to avoid the pitfalls of life in Inceville, with its muddy streets, roaming Indians, and wrong-side-of-the tracks ambiance. To this end it was announced in July, as if to prove that the movies had finally grown up, that "architecturally, it is intended that the new studios will be especially attractive."[5]

Ince also wisely configured departments that worked in proximity to one another in the same geographic area on the lot, so as to "minimize motion in handling scenery and players," as Ince's biographer would later describe it.[6] In addition to the usual studio departments, he included a general store, a hospital, and even a weather bureau to predict days when shooting outside would be impractical. He built a film lab that, unheard of at the time, could create release prints as well as the "dailies" used for film evaluation and editing during production.

The original plans called for two glass shooting stages, each 72 by 180 feet, although plans for a third one, the same size, were added soon thereafter. A backlot that was at the back edge of the property was also included. A 1922 article on the studio gives us one of the few written descriptions of this part of the plant during this time, stating "a number of permanent sets have been constructed on the lot, such as village street, fronts of buildings, the front and side elevations of fashionable homes and cottages of humble peasants. These sets can be dressed to fit the particular picture for which they are used, but the main parts of the structures are left intact, minor changes only being made."[7] This proto-backlot soon gave way to the more long-lasting second property across from the main lot.

The most striking feature of the studio was, as had been planned from the beginning, its administration building, a two-story structure with ornate, and in hindsight somewhat ironic, plantation-style columns. The building, known then and now as the Mansion, resembles George Washington's Mount Vernon when viewed across the lawn from the street, which, of course, is most significantly named Washington Boulevard.

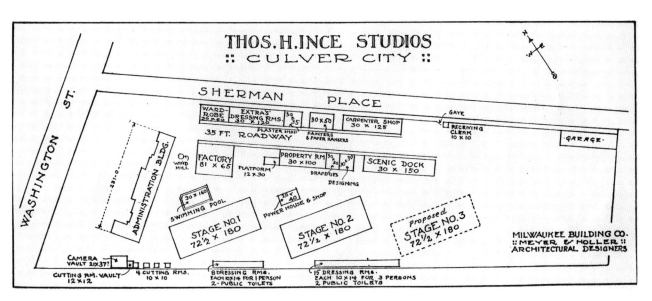

1918 studio plot plan

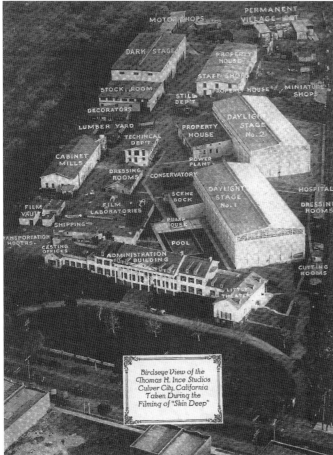

A 1922 aerial photograph of the lot and a promotional adaption of a similar photograph, most helpfully captioned with the assorted departments at work there during that era.

The Mansion, along with the rest of the studio, was designed in 1919 by the Milwaukee Building Company's Gabriel S. Meyer and Philip W. Holler, who in 1927 would design Hollywood's world-famous Grauman's Chinese Theatre (where the premiere attraction would be *The King of Kings* [PDC 1927], shot on the Ince lot!). The Mansion is sometimes identified as being built as a set for Ince's *Barbara Frietchie* (PDC 1924), although a look at the date for that production proves that the building was requisitioned, not created, for that Civil War–set production.

Ince's office, inside on the second floor, was adjacent to an elaborate dining room that was designed with a nautical flair, resembling, again somewhat ironically, a ship's cabin, which looked like the interior of his beloved yacht, the *Edris*. His love of the water was also reflected in a swimming pool, actually a studio tank, in back of the building—which gave the lot an almost jaunty, colonial revival / country club feeling, which was most incongruous indeed during this whirling makeshift era when most of his competitors' lots looked more like Inceville than Mount Vernon.

Harry Culver, obviously, was delighted to have Ince back. He showed his appreciation by throwing an elaborate party on January 4, 1919, which included a ceremonial "key to the city" presentation and an appearance by local balloonist Al Wilson, who wrote Ince's name in smoke above the studio.

Ince would run that studio from 1919 to 1924, continually reviving and streamlining both his production process and his production facility during his entire tenure, adding stages and sets and departments as well as genteel tree-lined sidewalks and landscaping between buildings, even as he added to his film library. He also, undoubtedly, made Harry Culver very happy when the King and Queen of Belgium,

Thomas Ince in his "ship's cabin" dining room, circa 1924, and the same location as it looked in 2013.

THE MAN WITH THE MEGAPHONE 11

along with Prince Leopold, toured the studio in 1919, the first ever visit by royalty to Culver City.

But Hollywood, or in this case Culver City, was changing in the 1920s. In hindsight, which is always the best way to view the rumblings of history, it is apparent that the very system that Ince had largely invented and streamlined eventually would have destroyed him as well. As an independent producer, even one with his own studio, Ince was, to a large degree, dependent on the majors for financing, talent, and distribution. By this time, he had moved into producing big-budget, prestige films like *Anna Christie* (Associated First National Pictures 1923), based on Eugene O'Neill's play, an interesting contrast to the costume-stealing Indians he had featured less than a decade earlier.

Like later independent producers such as Cecil B. DeMille and David O. Selznick, both of whom would inherit Ince's studio, Ince tried during this period to make fewer but better movies than the majors were able to create on their assembly lines—even though those assembly lines had been made possible through his innovations at Inceville and Culver City. But less product meant less profits for the producer, and so during these later years Ince often rented his studio's stages and sets to other independent productions, usually at a loss.

In the 1920s Ince rented his studio facilities out to other producers, a practice later owners of the lot would continue. (1924)

In 1919 Ince founded a company named Associated Producers with some other independents, but it was ultimately absorbed in 1921, somewhat ironically, by First National, which had been created with the same goals. Also, in 1921 he formed the Cinematic Finance Corporation, an attempt to secure financing for quality films produced independently. It too was ultimately unsuccessful.

In 1924 publisher-producer William Randolph Hearst entered into negotiations with Ince about using his studio as a base for his Cosmopolitan Productions. Cosmopolitan, like Ince, was an independent production company, but unlike Ince, it was exceedingly well financed, due to millionaire Hearst's personal involvement with star Marion Davies.

The original Ince backlot was in constant flux due to the very limited real estate available for standing sets. The second photo is from *Barbara Frietchie* (1924); the others have not been identified.

In November 1924, Ince boarded Hearst's yacht, the *Oneida*. Other guests on board that lost weekend included actors Davies, Charlie Chaplin, Margaret Livingston, and Jacqueline Logan; writers Elinor Glyn and (allegedly) Louella Parsons; Russian dancer Theodore Kosloff; and other Agatha Christie–style characters. At some point during the trip, Ince became ill and was rushed home, where he died at the age of forty-four of apparent heart failure.

Immediately, spurious rumors began circulating that Ince had been shot by Hearst, mistaking him for Chaplin after Chaplin had been caught in a dalliance with Davies. These rumors are apparently completely unfounded, and many authors and researchers and detectives have filled hundreds of pages of footnoted copy debunking them. Yet these same rumors, fanned by movies and books and speculation, have continued and have kept Ince's name alive and on the fringes of the public's interest long after his films and his innovations and his unfinished life have been, sadly, completely forgotten.

Today, all of the principals in this drama, even those possibly apocryphal Indians, have long paddled upstream into the dark compost of history. Ince too is gone, of course, like the letters of his name written in smoke on that long-ago opening day. Yet his studio—two of his three studios, in fact—survives today.

And in one of them is our story.

Thomas Ince promotional emblem. (1920)

CHAPTER 2
Moguls, Millionaires...
and a Bandleader

You won't find that island on any chart.
—Carl Denham, *King Kong*

Cecil B. DeMille

Thomas Ince's widow, Elinor, held onto the studio property until February 1925, when it was sold to Ince's friend, Cecil B. DeMille. DeMille had recently broken with Paramount, a studio he had helped found, and was striking out as an independent. Partnered with a devoutly religious banker named Jeremiah Milbank, and with $50,000 worth of lighting and camera equipment given to him by Paramount as part of his separation agreement, DeMille started a short-lived company named Producers Distributing Corporation (PDC). A third partner quickly came to the table as well: Pathé, an already venerable production company, which would

Cecil B. DeMille—center, of course—in front of his new studio, April 15, 1925.

ultimately provide most of the money, as well as the distribution pipeline promised by the name. DeMille and Milbank, or rather, Pathé, paid $500,000 for the Culver City property, which he renamed Cecil B. DeMille Studios.

Part of the deal with Mrs. Ince was that the Mount Vernon facade would remain intact—which was fine with DeMille, who did add other buildings behind the Mansion. He also, like Ince, brought in independent producers to rent some of those buildings. A telegram exists in the collection of Bison Archives indicating that DeMille, in 1924, tried to talk his old Paramount colleague, Samuel Goldwyn, who had himself recently struck out as an independent, into joining him in Culver City as a tenant. Goldwyn declined. DeMille, apparently not superstitious, about the same time also installed a high-tech-for-its-time radio transmitter-receiver so that he could keep in touch with the studio when aboard his own yacht, the *Seawind*.

In February 1926 DeMille signed a $10,000, four-year lease on the property behind the studio, which would be known as 40 Acres. The *Los Angeles Times* reported, "This land, owned by Achille Casserini, adjoins the property of the studio on the east, and includes a stream and numerous large trees which will be used to advantage in the construction of exterior settings on this added acreage."[1]

Interestingly, the acreage of the property was here listed as being 42 acres. Confusingly, two years later *Variety* attributed the property as being both 62 and 67 acres—in the same article.

DeMille also inquired as to an additional 56 acres of adjacent property at the same time also owned by the Casserini family, but that deal was never completed. We do know that in 1928, one H. G. Phillipps was involved in a legal action against the Casserinis regarding legal ownership of the 40 Acres property, which he (Phillipps) ultimately lost. Whether or not this question as to the physical ownership of the property, or DeMille's pending exit from the company, had anything to do with his ultimate failure to carry out these ambitious plans is unknown.

Jake Shepherd, in his study of 40 Acres, has observed that "there has been controversy concerning the amount of acreage on the backlot, primarily because of the existence of a May 9, 1940 plot that shows the

A heavily retouched Harry Culver and his wife, Lillian, give Cecil B. DeMille (center) a cheap-looking ceremonial key to welcome him to Culver City. (1925)

DeMille's studio in 1927. Note the sets at the back of the main lot and the *King of Kings* structures on 40 Acres.

backlot portion on the north side of Ballona Creek to be roughly 26 acres and the portion south of Ballona Creek to be roughly 10 acres."[2] Shepherd believes that the original size of the property was indeed 42 acres, but that in 1935 a Ballona Creek flood caused the Army Corps of Engineers to rebuild the creek and the creek bed, which seems to have dissected a few acres off of the southern part of 40 Acres. The construction of Jefferson Boulevard, which did not really exist when the property was first leased, could have also robbed the property of some of its acreage. These "improvements" carried on intermittently until 1938. The final size of the backlot eventually ended up as being about 29 acres.

A 1932 lease exists between what was by then RKO, or RKO-Pathé Studios, and Archille Casserini regarding the lease of 40 Acres to the studio for one year for $6,000. An interesting, nonsensical clause in the document states that "leasee may build and use motion picture sets, which must be removed on expiration of term." Obviously, as removal of valuable standing sets before this expiration of term would be detrimental to the property's continuing use as a viable backlot.

The truth is that the sets at 40 Acres were *always* in flux, always changing, always mutating into something else, as somewhere else. A 1931–32 inventory of the property includes, just as a sample, the following sets and buildings on the lot. Some of these names will be familiar after reading Chapter 4; others' origins and usage, and ultimate fate, will probably always remain a mystery. The sets with dates cryptically listed after the set name indicate that that set was constructed, removed, or perhaps scheduled to be removed during that time, in order to build heaven knows what:

"Castle," "French Street," "Log Cabin (10/3/31 to 1/2/32)," "Dutch House and Barn," "Farm House and Barn," "Miniature Roof (10/3/31 to 1/2/32)," "Bridge (5/30/31)," "Belgian House," "Old Mill and Stream," "French Church (10/3/31 to 1/2/32)," "French House and Barn," "Barn," "German House," "Dutch Tavern," "Combination House and Jail," "English House," "Warehouse and Meat House," "Biblical Castle," "German Village," "French War Village," "English House," "France [sic] House—two story," "Arabian Court," "Egyptian Court of Castle," "Adobe Indian Hut," "Spanish Courtyard," "Ship Deck (10/3/31 to 1/2/32)," "5th Avenue Residence," "Chicago Street," "State Reform School," "Generator, Havana Street Set (7/4/31)," "Ext. Golden Arms (7/4/31)," "Ext. Temple and Gardens (4/4 to 8/1/31)," "Ext. Pool Room and Café (NY street; 8/29 to 10/3)," "Ext. Courtyard (8/29 to 10/3)," "Submarine Enclosure (again, 8/29 to 10/3)," "Jungle Set (10/3/31 to 1/2/32)," and "Auto Court and Gas Station (10/3/31 to 1/2/32)"—remember, all of this was during an approximately twelve-month period in the early 1930s.

A 1932–33 lease renewal agreement between RKO and the Casserini family for 40 Acres. The studio would not actually purchase the property until 1946.

And yet, in spite of the backlot's seemingly constant and continuous usage, it was not until 1946, after years of haggling and threats and last-minute lease renewals between the Casserinis and RKO, that the property, including what was always referred to as the "Old Orchard" across Ballona Creek, was finally purchased by RKO for $152,000.

As these sometimes-conflicting documents prove, the name "40 Acres," in spite of all this interesting minutia, was probably never intended to be taken literally. The term actually refers, somewhat sarcastically, to the back side of the farm, or the "back 40," deeded to settlers in the federal Homestead Act of 1862 and its various amendments, which opened up the American West to farmers and pioneers. Likewise, 40 Acres, as used here, was the back of the studio "farm," or the back "lot" of the property. In *Gone with the Wind* (MGM 1939), it should be noted, there is also a scene, set during Reconstruction and shot on 40 Acres,

18 HOLLYWOOD'S LOST BACKLOT

DeMille's great white hunter–inspired office on the third floor of the Mansion reflected a longing for a different sort of adolescent adventure than Ince's dining room had. (1927)

where the line "We're going to give every last one of you forty acres and a mule" is evoked.

Finally, for those of a literal frame of mind, the total acreage of both studio properties, front lot and backlot, was approximately, yes, 40 acres.

DeMille's office was on the third floor of the Mansion (writers and director were on the second floor) and was even appointed with a white bearskin rug. The studio's research department was next door, although most of the books there were from DeMille's personal library. Outside, on the driveway, a uniformed butler, the impressively monikered Hillard Conwell, stood by to open the Mansion's white paneled doors for guests.

As a director, DeMille's big on-screen opus while on the lot was nothing less than the life of Christ, *The King of Kings*, which his pious partner Milbank, in particular, was keen on, although it must be said that little else about Hollywood impressed him. That film was wildly successful (as of 1935 it had grossed $2,602,272.52), although its scope and expense, and a most generous handout to charities and churches, ultimately limited the profits. Other DeMille pictures, especially the ones he did not direct himself, were less successful. The least successful

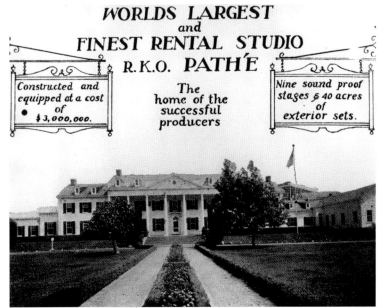

Like Ince and DeMille, RKO-Pathé also attempted to enhance its bottom line by renting studio space out to independent producers, as illustrated by these 1931 trade magazine ads.

picture during DeMille's reign, however, was one of his own, *The Godless Girl* (Pathé 1929), which lost $233,000.

RKO

In 1927 Joseph P. Kennedy, who had recently acquired, through takeovers, the Keith-Albee-Orpheum theater chain, and the FBO Pictures Corporation—which itself was partially owned by radio manufacturer RCA—purchased Pathé as well. DeMille, realizing there could be room for only one autocrat on the lot, and also realizing it would not be him, sold his interest in the company, which changed the name of that company, and that studio, to Pathé.

Kennedy, who was disdainful of Hollywood in general and who found DeMille in particular a bit of a martinet, eventually brought most of the production and exhibition companies he had captured together as a new major, RKO (Radio-Keith-Orpheum), in October 1928. The Pathé Studio was largely outside the parameters of the original deal but was purchased separately by RKO, for $4,630,789, in January 1931.

According to historian Richard B. Jewell, the studio in Culver City was one of the factors that made the company attractive, as that studio's Gower Street lot (in Hollywood) was "small, with very little room for expansion or the erection of exterior sets, the additional studio space would be a welcome addition."[3] This latest maneuver changed the name on the sign outside the Mansion once again, this time to RKO-Pathé. DeMille, probably happy to not have to worry about the nine hundred employees on his studio payroll anymore, moved a few blocks west on Washington Boulevard to MGM and eventually back to his old studio, Paramount, where he had started and which he had started, and where he would remain for the rest of his long career. Kennedy's Hollywood period, by contrast, was to be short-lived. "Joe Kennedy had no particular interest in the movies, beyond functioning as a quit-claim operator and sleeping with Gloria Swanson," Scott Eyman, one of DeMille's biographers, once acidly commented.[4] Only the Bible-clutching Jeremiah Milbank, whom DeMille somehow remained on friendly terms with, got out of the movie business altogether and on his own terms. In February 1932 RKO officially eliminated the Pathé name, merging the company with its Radio Pictures unit. It then closed the so-named studio for reorganization, opening it again in May of that year. The name of the lot would continue to be Pathé, however, in public and private conversation and paperwork in order to avoid confusion with the "other" RKO studio on Gower.

The name RKO-Pathé would not last long, although "Pathé" would be used to describe the studio for the next quarter century. (1930)

MOGULS, MILLIONAIRES . . . AND A BANDLEADER

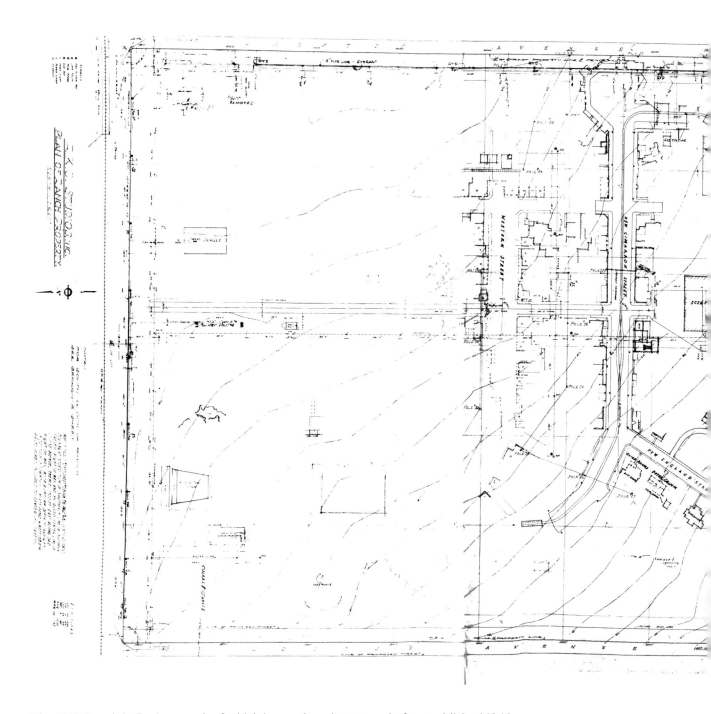

The RKO Ranch in Encino, much of which is seen here in a never before published 1940 map, competed with 40 Acres for the attention of both RKO and independent producers for nearly thirty years. In spite of the ranch's enormous size and the vast variety of its sets, 40 Acres would ultimately win that battle.

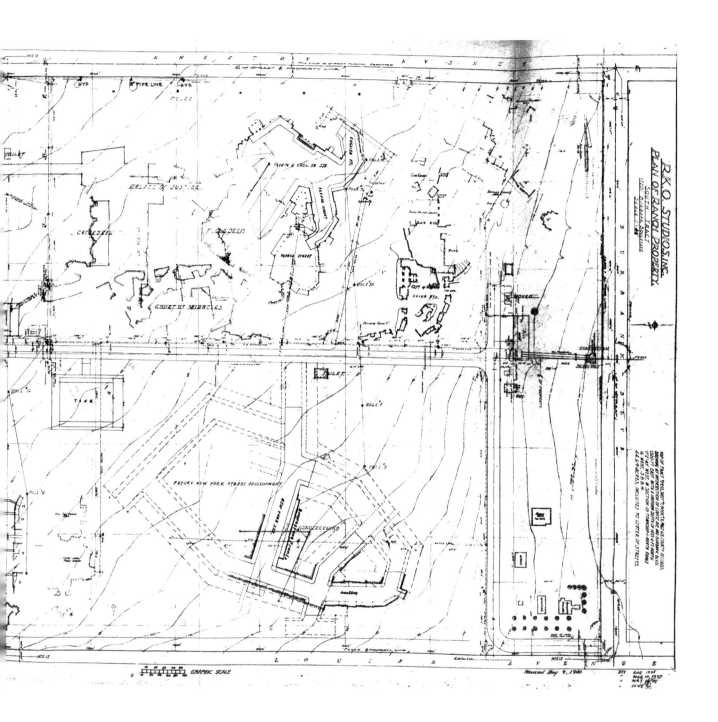

RKO's management of the lot would last for decades and should seemingly have provided the stability for the property that it had lacked for its first chaotic years. But this was largely not to be, as RKO would prove itself to be highly unstable, with a revolving-door procession of management personnel parading through those doors throughout the company's chaotic and colorful history. In fact, in spite of the Culver City lot's perceived attractiveness, RKO management seemed to be somewhat averse to shooting in Culver City at all. Although hundreds of famous and forgotten RKO pictures would be shot in whole or part at Pathé, RKO often rented the studio out to other studios and to independent producers, ultimately operating more as a landlord than as a tenant.

Additionally, in 1929 the studio leased, and later (in 1939) purchased, an (initially) 500-acre "ranch" in Encino, in the San Fernando Valley, where huge sets could and were constructed as well. However, this further decentralized the company. In 1935 studio general manager J. R. McDonough in a memo to one M. H. Aylesworth complained that "we have been renting the Pathé facility to independent producers because it is so far distant from this studio that we cannot afford to use it unless we set up separate production units there. Our own ranch is not used much because of its great distance from our studio and also because of its lack of dressing room facilities, restaurant facilities, etc. The result is that we are confined largely to the use of this Gower Street studio as to RKO productions."[5]

McDonough was trying to acquire funding to build new stages on Gower Street at the time he authored this memo, so he probably purposely overstated how distant Culver City was from Hollywood (for the record, the Hollywood lot was, and is, 8 miles from Pathé). But it is true that a certain lack of momentum—which is evident in the RKO production records, with their somewhat exasperated reports of travel delays and forced company moves—was apparent in any production trying to shoot in both, or all three, "RKOs."

No one ever seems to have considered that it would have been more practical to have spent the $200,000 McDonough wanted to improve the infrastructure at Pathé, which already had an extensive backlot, rather than at Gower, where nearly all exteriors would *always* have to be shot off-lot, no matter which improvements were implemented on that lot, due to the space issues. In fact, the Culver City lot then had eleven stages, and Hollywood only had ten. McDonough would get his three additional stages at Gower at the end of that year, which along with other improvements would end up costing $500,000, more than twice his original request.

David O. Selznick

In 1935 RKO-Pathé, including 40 Acres, was leased by David O. Selznick. *Variety* reported that year that Selznick was paying $125,000 annually for the studio and an additional $4,292.50 for the backlot, although RKO would continue to use the property during this period as well. A 1938 internal memo placed the annual rental for both lots at $114,292.32.

Like previous tenants Ince and DeMille, Selznick was a formidable independent producer looking for a base of operations. Unlike Ince, who had remained an outsider, even as he had created the patterns and prototypes for the studio system, and DeMille, who was the ultimate insider and therefore uncomfortable on his own, Selznick was both an insider—he was, after all, married to MGM's Louis B. Mayer's daughter Irene and had been an executive at Paramount, MGM, and RKO—and yet, on a bedrock level, he was also a rebel.

Selznick's father, Lewis J. Selznick, had been one of the pioneers of the motion picture industry in its earliest, East Coast era. Realizing that the industry was moving west, the senior Selznick had tried to move his operation to Hollywood in the early 1920s but had been destroyed by the major studios that preceded him. By the time he died in Hollywood in 1933, his older son, Myron, had grown up to become an agent, practically creating the blueprint for the ruthless Hollywood "ten-percenters" who would follow in his contentious wake and extort the coffers of the mighty empires that had destroyed his father. David, the younger son, was determined to work not from without, like his brother, but from within—from the other side, from the inside of those empires. David's revenge for the sins against his father was to restore the Selznick name to its former prominence in the industry, from within that industry.

As Selznick worked his way up through the ranks, one hated major at a time, he realized that the way to achieve this revenge was not to be *at* a studio at all, but rather to *be* a studio. And unlike his carpetbagger brother, he longed to accomplish this all from within that industry. It was therefore vitally important for Selznick to have his family name not just in, or on, the movies but on the gates of an actual movie lot.

The Ince-DeMille-Pathé-RKO lot was ideal for this purpose. As mentioned earlier, RKO, somewhat inexplicably, wasn't sure what to do with the property, yet it was already a ready-made, self-contained film factory. In fact, the lot was the prototype for what a film factory should be. So, it didn't have to be reconstructed or reengineered, rejiggered, or brought, to use a crass contemporary term, "on-line." It just had to be occupied—the switch just had to be thrown. All that really needed to be done was to paint a white sign to put in front of

The Mansion was never owned by David O. Selznick, nor could the company he founded be legitimately, honestly, ever be called "International." Yet to this day it is his name the public remembers on this driveway. (1935)

the Mansion. So that sign, that mansion, would soon become the logo for Selznick International, which was the name of the company Selznick would create.

Selznick, in an uncredited and undocumented story, which we will repeat here anyway, reportedly called his wife, Irene, in 1936 and told her, "Drop everything you're doing and come to the studio—at once!" Mystified, Irene left her home and drove swiftly to Culver City. Turning off Washington Boulevard, she motored up the curving driveway to the studio. David was standing in front. Irene leaped out of the car and asked him, "What is it, David?" He grinned and gazed up at a painter on a scaffold above the gleaming white colonnade. The painter was finishing a sign that read "SELZNICK INTERNATIONAL STUDIO." David and Irene embraced each other and cried.

It's a good story.

Selznick released his first two movies, *Little Lord Fauntleroy* (United Artists 1936) and *The Garden of Allah* (United Artists 1936) without any company logo, which he considered a wasted opportunity to build up trademark value. But during the production of *A Star Is Born* (United Artists 1937), Hal Kern, Selznick's favorite editor, drove through the front gates of the lot and noticed how visually striking the Mansion was:

> There were some beautiful white billowy clouds behind the building, the lawns were a bright green, and the building itself was very imposing. So, I thought we should photograph this and use it until we got a permanent trademark. So I got Duke [Howard] Greene, the Technicolor cameraman, and we took the camera out just a little after noon, and photographed the sign close up and then swung down for a long shot of the building, and just as we did a flock of birds flew over the main building and some guy came out the front door, but we used it anyway, because of the birds."[6]

That image—although it would be reshot several times, with the later addition of composer Alfred Newman's stirring musical fanfare and the motto "In a Tradition of Quality," a tribute to his father—would be used by Selznick as his trademark for the rest of his life.

Selznick broke with tradition somewhat in that his office was not in the Mansion at all, but rather in an annex to the east of the Mansion. Inside this office is where Selznick stirred deals and decisions and Benzedrine and business, making movies and mayhem, and where, in marathon sessions with screenwriter Ben Hecht and director George Cukor, much of the screenplay for *Gone with the Wind* was forged. Even today, against all odds and all common reason, outside of the windows of Selznick's old office—tangling itself around a garden-hose holster, engulfing the rest of the landscaping, and threatening to choke out the delicate and carefully landscaped flowers—resides a mint plant, placed there all those years ago, and all

David O. Selznick's office, as seen here in 1940, was subtler and more sophisticated than the offices of Ince and DeMille, as second-generation mogul Selznick had the advantage of a better education and more refined tastes than his predecessors. His office isn't as interesting though, is it? Inset: The preserved sign that hung outside Selznick's office as it looked in 1989.

those movies ago, so that the mint juleps in that particular movie would have their primary ingredient readily available on the lot.

Selznick, unlike RKO, appreciated the ambiance and tenor of his new studio. He liked to roam its streets and was often photographed looking reflectively across the sets out on 40 Acres. He probably enjoyed greatly an inaccurate but (more importantly) impassioned 1936 *Los Angeles Times* article stating that "next to Universal, the eighteen-year-old RKO Pathé 'back ranch,' also known as 'the Forty Acres,' probably is more seeped in cinema lore than any other. A leased stretch back of Culver City, it served Thomas H. Ince in the Triangle days, and later Cecil B. DeMille."[7] Again, it's a good story.

The studio lot, and backlot, was indeed an evocative, even romantic, place by this time. Quite unlike any other factory in Hollywood, this studio had a peculiarly genteel, almost Southern quality, with well-landscaped hedges and white clapboard bungalows situated on little lawns at the end of winding sidewalks—all of which suited and nurtured Selznick's romantic sensibility, and the romantic sensibility of the movies he would make there. The resentment and the admiration, the longing

David O. Selznick liked to walk the lonely streets of his backlot for inspiration. The partially obscured train funnel in the foreground, like all of the buildings, is artificial. (1939)

to avenge his father and then to surpass him, to destroy his father's destroyers and then to surpass them in artistry and success as well, were the wellspring of his career.

Selznick would eventually succeed in all of his goals. But he would pay a price.

Actually, during the late silent era, there had been a certain almost madcap, garden-party ambiance about filmmaking that had been reflected in the architecture at several of the studios, which often sprouted gardens and topiaries amidst the sets and stages. But with the coming of sound, and the heavy industry that sound technology heralded, the lots had become much more industrialized. Even the majestic MGM, just up the road, had, by 1935, acquired a well-oiled mechanized ambiance, which belied the ultimate glamour of their product.

This was not the case here, however, because the previous owner and landlord, RKO, had never bothered, or had never been able to afford, to do more than the barest possible modifications on the lot. In fact, decades after the coming of sound, in 1964, angry memos during the production of the *Star Trek* (TV 1966–69) pilot "The Cage" complained about the bad soundproofing in Culver City.

Also unusual was the physical shape of the studio, because of its various contractions and reductions and because of the unusual terrain in the neighborhood,

28   HOLLYWOOD'S LOST BACKLOT

An aerial view of the lot well illustrates the studio's unusual physical properties. 40 Acres is top, left. (1939)

namely the nearby Baldwin Hills and Ballona Creek. The studio proper looked, by this time, like a lopsided triangle-trapezoid that gets narrower, to the frustration of truck drivers trying to navigate its streets, as it staggers toward the hills. 40 Acres, which almost, but not quite, touched the main studio at the intersection of Lucerne Avenue and Ince Boulevard, was also pie-slice-shaped. So, the two properties when viewed together looked a little bit like a gardener's hoe, or the head and neck of an inverted pelican, or a broken arrow, or a badly knotted tie, or a blurred wheel spinning in a movie montage.

Culver City itself, by its very nature, is one of the most confusing neighborhoods in Los Angeles. There seems, there really does, to be no true north anywhere in the city. The whole place is laid out in Dutch angles. Naomi Hirahara, a popular Southern California author, once memorably, and accurately, said about Culver City that "a lot of its streets tangled up in knots like the roots of a tree smashed into a pot that was too small."[8]

Yet Selznick, the impractical romantic, loved the place. He built out the bungalows and landscaped the lawns and painted the buildings. He, or rather RKO, also added two more stages (15 and 16) in 1940. And so, many movies and many years later, when it was all over and was all being liquated, it was the loss of his studio, even more than his stars and his company, that the great producer regretted the most.

In 1938 there was a bizarre and tragic incident in the so-called Arab Village backlot on 40 Acres. A 13-year-old boy, James Frantz, was shot in the abdomen by policemen who were investigating a string of recent burglaries. The two police officers said that when Frantz broke from cover and attempted to flee down the stairway of a "mosque" set, the boy was shot by mistake. The Los Angeles Times reported that "while awaiting the ambulance the boy assertedly confessed to 15 burglaries. Police said he was committed to Whittier State Reform School last Aug. 31, escaped and was recaptured on Oct. 27, and made another escape four days later."[9] Whether or not the young delinquent was actually *living* in these sets while a fugitive was not discussed in the article, and no word about his ultimate fate was ever reported.

During his years on the lot, Selznick's relationship with his landlord, RKO, was sometimes contentious. A consistent issue between the two companies seems to have been over the usage of the backlot. For example, for *Gone with the Wind* an antebellum Atlanta, Georgia, was built on top of, and out of, an older "Chicago" set from the DeMille era. Naturally, RKO argued, in endless memos and lawyer letters, that this newer set was not as viable/rentable/usable for future productions as was

Gone with the Wind art director Lyle Wheeler plans vast changes to the 40 Acres backlot using a detailed scale model depicting the many sets to be built or adapted for that picture. Landlord RKO would not be pleased. (1939)

the previous version. An undated (circa 1942) RKO memo, for example, complains about "Selznick Pictures covering our old concrete street with approximately 2 feet of dirt," and bemoans the $875 material and $925 labor costs—and the 27.5 percent overhead that would be required to set it all right again.

In an internal 1941 memo, Selznick's production manager, Ray Klune, admitted that RKO's claim was to a small degree a valid one, but apparently afraid of telling his mercurial boss something he didn't want to hear, Klune then argued that "with minor exceptions and one major exception (Tara) most of the sets and set elements left here by us have already lost their identity through the normal evolution of sets."

In 1938, Selznick in one of his famous memos, likewise questioned the logic of building and paying for sets on a property he didn't own: "I also wonder whether we have given any thought to the possible advisability of buying a piece of property immediately adjacent to here, so that we could own the set. The property would presumably be worth what we paid for it, or somewhere near it—quite apart from the set—and the set ought to bring us steady revenue for a period of years. It does seem wrong to invest, perhaps $50,000 in a street for the benefit of RKO."

These exchanges were typical. File cabinets at the RKO and Selznick studios eventually filled up and then overflowed with notices about sets being built or demolished at 40 Acres—and who was to ultimately pay for those venerable sets' construction, upkeep, and, finally, their demolition.

Selznick, like RKO, did sublease these sets out to third-party clients during his tenancy on the lot. In 1938, for example, a typical set at 40 Acres could be had for between $300 and $350 a day, with a 10 percent restoration fee added. Selznick's camera department would also give the client a Mitchell 35mm movie camera for an additional $15 with a set rental.

Sometimes these same sets were also destroyed unintentionally by one party or another, or by one of those third parties, or even by an act of nature. The first recorded fire on the lot was reported in 1927 when Culver City firemen responded to a "stage and set" fire at the DeMille Studios, which destroyed Stage 6. A 1936 blaze destroyed the old "South Seas Set," which probably referred to some of those constructed for *Bird of Paradise* (RKO 1932). In 1948 there was a projection room fire on the lot, and in 1959, after RKO was defunct, unknown footage they had left behind self-combusted in a studio vault.

In 1942, during production of a United Artists picture, *The Moon and Sixpence*, which was renting space at 40 Acres from either Selznick or RKO, an on-screen fire turned into the real thing when flames from a Polynesian-style grass hut, which had been fitted with gas lines in order to burn on cue, leaped over to some adjacent mock palm foliage. Fortunately, studio firemen quickly put out the blaze.

A year later another fire occurred, this one possibly caused by some boys "who were seen playing in the vicinity of the sets shortly before the flames erupted," as the *Los Angeles Times* reported.[10] The *Times* also curiously noted in their coverage of

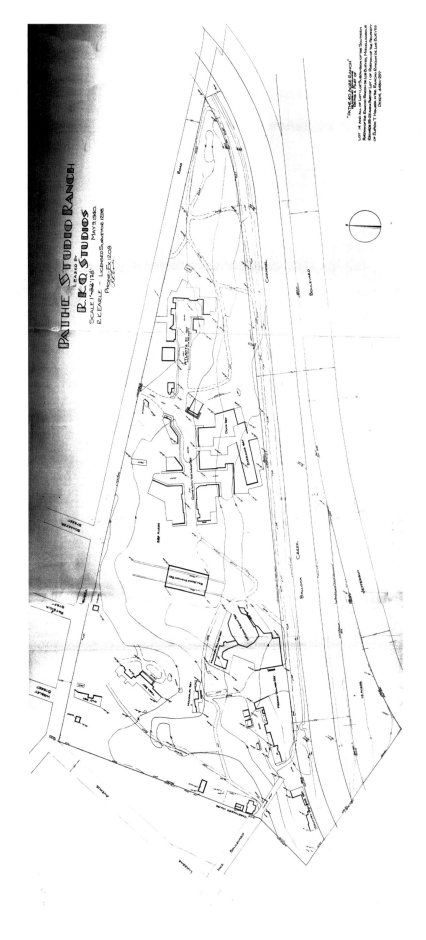

40 Acres in 1940 included sets intriguingly left over from the silent era and from several then-current Selznick and RKO productions. Many of these structures would be short-lived, yet a few would remain on the lot for the next thirty-five years.

the fire that "a courtroom set used in the picture 'Gone with the Wind'" was one of the sets razed by the flames[11]—although among the hundreds of sets constructed (or according to RKO, ruined) for that movie, a courtroom setting was never one of them.

In 1946 *Variety* reported yet another fire at 40 Acres, which this time reportedly destroyed the Aunt Pittypat house from the same film, although the set, or something that looked like it, would decades later make postmortem appearances on TV episodes of *The Untouchables* (1959–63), *Land of the Giants* (1968–70), and *The Andy Griffith Show* (1960–68).

Selznick International Pictures would suspend operations in 1943, although Selznick would continue to use the name publicly as part of a partnership deal with United Artists. Vanguard Films would be the new company name, at least on paper, until 1951.

In 1947 Selznick, who less than ten years earlier had produced the most successful movie ever made, was forced to give up part of his lease on his 40 Acres empire. Although he would continue to maintain studio and office space on the lot, one expensive and unsuccessful film after another, as well as his impossible obsession with "topping" *Gone with the Wind*, had crippled the once-mighty producer financially to the point where he could no longer afford most of his own studio. An ill-advised 1946 decision to release his own product, through the short-lived "Selznick Releasing Organization" (SRO), did not help things.

In 1949 auctioneer David Weitz held an auction of Selznick's assets on three of the soundstages on the lot. Among the items for sale were "lumber, building equipment, cameras, office furniture, antiques, wardrobe, make-up, greens, 12 autos, sound and grip equipment among other items."[12] The *Hollywood Reporter* noted that ads for the auction would not carry Selznick's name, presumably to avoid embarrassing the producer. The sale was a grisly prelude to later, even more extensive auctions by the studios of their heirlooms, including a now-legendary public bloodbath conducted, ironically again by Weitz, at MGM in 1970.

So Selznick was probably referring to his studio, at least in part, when he confided in 1954 to his friend Ben Hecht that "Hollywood's like Egypt . . . full of crumbled pyramids. It'll never come back. It'll just keep on crumbling until finally the wind blows the last studio prop across the sands."[13]

In 1952 Louis B. Mayer, Selznick's former father-in-law (David had divorced Irene in 1948 and married actress Jennifer Jones), asked if he could borrow Selznick's remaining offices at the "Selznick" Studio. Mayer had recently (in June of 1951) been forced to resign from MGM, and now he needed a place to work from. According to his biographer, David Thompson, Selznick responded that:

"It distresses me to have to advise you that I really don't think it will work . . . I have paid a fortune in rent to maintain these offices for a period of years during which I have been there very little." Selznick went on to explain that the offices were crammed with files, "trick equipment, reversible projection equipment, etc." and that they were "terribly run down." He concluded by admitting that "I myself feel, when I go there as though I were living somewhere in the broken-down remnants of the Old South; but I was there in the studio's better days, and it isn't the same thing as your moving in for the first time."[14]

Mayer, presumably still officeless, would die in 1957.

In 1955 Selznick's name would again go up in front of the Mansion, briefly, when he partnered with RKO as a producer for an "indeterminate" number of films, which, unfortunately, would turn out to be zero. RKO had just been purchased and was itself in a financial freefall at the time—and it was likely that the new owner, Thomas F. O'Neil, only hired the somewhat desperate Selznick for his name and for access to the Selznick film library, which they hoped to rerelease, although, in truth, much of that library had already been sold. The announcement came just after the producer had vacated a post-Mayer MGM—where again, he had not been able to get a single picture past the screenplay stage.

Two years later, however, Selznick would manage to mount one last film, *A Farewell to Arms* (1957), for yet another studio, 20th Century Fox. The picture would open not with the Fox fanfare, however, but rather, and for the last time on screen, with that beautiful shot of the Selznick sign and that white mansion receding in the background. Selznick would die in 1965.

Back on Gower Street, RKO was purchased in 1948 by billionaire Howard Hughes. In print, Hughes's name is usually preceded by the word "eccentric." True to form, the industrialist's (mis)management of the studio was an unmitigated and legendary disaster. Hughes tinkered with his producers and his productions, sometimes delaying those productions' release for years. He interfered with his directors, his crews, and his executives. He also legendarily interfered, or fired, or mismanaged, or dated the studio's stars. Obviously, this tinkering and second-guessing dramatically affected the always beleaguered studio's plunging bottom line. In 1949 RKO's profits freefell a quite remarkable 90 percent. Writer Betty Lasky once sarcastically, but not untruthfully, called Hughes's involvement with the company a "systematic, seven-year rape."[15]

In 1953 the by then only 89-acre RKO Ranch in Encino was liquidated for a meager $537,500. Yet surprisingly, the Culver City properties were not parceled off

David O. Selznick promotional emblem. (1957)

at the same time. Instead, in 1955 the perpetually lucky Hughes sold the lot, along with the Gower Street plant and the rest of RKO's assets, to businessman Thomas F. O'Neil's General Tire and Rubber for $25 million. Hughes realized a nearly $7 million profit on the deal—but handed over a studio tied in chaotic knots, which the new owners, who had hoped for an entrée into the entertainment business, were unable to untangle.

Desilu

In February 1957 Selznick's old production manager, Ray Klune, now floundering about at RKO, recommended that the company "dispose of the Culver City property at the earliest possible moment."[16] John Wayne and John Ford were rumored to be interested, yet it was not until November of that year that Desilu, the television production company owned by former RKO starlet Lucille Ball and her *I Love Lucy* (TV 1951–57) costar and husband, bandleader Desi Arnaz, bought both studios (and 40 Acres) for $6,150,000, although escrow would not close until January 25, 1958, when Desilu took possession of the property.

The Mansion in 1958, shortly after Desi Arnaz and Lucille Ball took possession of the property.

During a period when "movie" studios had, until recently, been resistant to television and were selling off their backlots, Lucy and Desi were delighted by their purchase, and never understood why television and films could not work side by side on the same property at the same time, which today is what happens all the time. The couple also, in some ways, supplemented and improved the property, with new sets and additions to facades that had already been standing on the backlot for decades. In 1961 when real estate developers tried to buy 40 Acres for a shopping center, a spokesperson for Desi flatly stated that "we're not interested in making lands deals."[17]

Desilu also always featured 40 Acres prominently in promotional materials and publicity. During a 1957 helicopter flyover of the property for Westinghouse, one of their sponsors, Desi, looking down into the seemingly endless facades, couldn't resist rhapsodizing like an excited Cuban schoolboy about how "all this stuff is for exteriors, outdoor action stuff. Right now, we're building a whole village down there! A whole western town!" Almost as if he couldn't believe his luck, he added that "to give you an idea about how big it is, that's where they made *Gone with the Wind*!"[18]

In 1959 Desilu did allow the Continental Oil Company to look for petroleum in Culver City. MGM signed a lease with the same company at the same time to do the same sort of exploratory drilling on their lot. Eventually Conoco did sink a

Continental Oil sank exploratory oil wells on 40 Acres, but the big one never came in. (1963)

Desi Arnaz (in hat) surveys his tattered yet romantic empire. (1958)

well at 40 Acres in back of *The Real McCoys* (TV 1957–63) sets along Ballona Creek but the Big One, at least not in oil, never came in.

Surprisingly, it was during the Desilu era that 40 Acres came into its own in a way that had not really happened during any of the many other regimes. The Arnazes had made their fortune in television, and the backlot proved to be a valuable asset in producing for that medium. The life of the property was extended by more than a decade by both Desilu's own series and by other shows and feature films that rented the sets there. Between those sets, and the thirty-three combined stages in Culver City and Hollywood, the company became, as *Variety* put it in 1963, "the largest indie rental studio in the industry."[19]

This idea of renting facilities to outside clients was not a new one. Rental studios, sometimes called producer's studios, had long operated on the edges of corporate Hollywood, leasing their facilities to independent producers, or to the majors when they got lucky enough to do so. Additionally, those majors had always rented, or bartered, or borrowed sets and stages and equipment from each other, as their production schedules and needs demanded. But in the 1960s studio income was down and studio overhead was up, and Desilu, probably Desi—who was proving to be a very innovative executive—saw that vast amounts of money could be made by renting facilities to the majors and doing it more economically than the majors could themselves, even on their own properties.

It all worked remarkably well, with the studio often running at capacity. By 1962 the loans on the properties were paid off—two full years ahead of schedule. *Daily Variety* characteristically reported that "the lot is SRO."[20]

In 1963 Desilu had 130 permanent employees, as opposed, for example, to the 3,500 on staff at MGM. At the time only three of the thirteen series shot on the lot were Desilu-owned. The others were tenants—resulting in billings for the company that year of $22 million.

Standing behind Desi on the sidelines, Argyle Nelson, Desilu's VP in charge of production, marveled that "the sound department doesn't cost us a cent. Glenn Glenn [sic] handles the sound and we participate in the gross. We pick the best outside jobbers to take care of the tenants. We have an independent casting operation. When we are not in production we don't even have assistant department heads on the payroll—just department heads."[21]

CHAPTER 3
Front Lot: Sets, Settings, and Set Pieces

> Perhaps the most exciting part of a visit to the Ince studio is that moment when awareness floods in—in some magical way you are standing on a patch of Hollywood's "sacred ground." As you stand in that damp, drafty, historic structure known as Stage 1, examining the windows from Scarlett O'Hara's bedroom, touching the English phone booth used in *Rebecca*, and inspecting the train station from *Spellbound*—in that instant, the romance, myth, and fantasy of the studio fused with its pragmatism, its history, and its reality.[1]
>
> —Isobel O'Neill and Dana Privitt

COMMENTS, EXPLANATIONS, and excuses are in order here. Every attempt has been made to verify every title mentioned in connection with every stage and every set explored in this and the following chapter. Much of the information has come from careful, sometimes frame-by-frame viewings of many of these films, which can be problematic because it is sometimes rather difficult to verify a backlot set's often heavily camouflaged appearances. And, of course, soundstage-wise, a viewing of a film yields few answers anyway.

Some titles referenced here were also verified through production records, which is fine and well and generally a very trusted resource in regard to those soundstage appearances. Out on 40 Acres, however, these same records have proved themselves to be considerably less reliable. Oftentimes the description listed in the daily production reports would be "40 Acres, tenement window" or something equally vague. In some cases a careful, if guilt-ridden, guess has been made. A tenement would only

The Mansion is timeless. Only the names above the door have changed since 1918. It was not given official landmark status, however, until 1990. (1970, Mansion photo courtesy of Rex McGee.)

be found on a city street set, right? But the original descriptions have, whenever practical, been preserved between quotation marks, so other interpretations are, of course, welcome—unless these guesses disagree with ours.

Unlike other "studios," this lot in Culver City was never the central headquarters of any of the seven majors we think of in regard to Hollywood. Yet all seven of those majors used the lot, and one of them, RKO, shot pieces of hundreds of their releases there, even while others were filming around them at the same time. So, for the sake of clarity, the original distributor as well as release date, which can differ from the production date, has been included on feature titles throughout the book, at least upon a film's first mention. For television productions, increasingly important to the lot from the 1950s on, only the designation "TV" and the original broadcast dates are reported here.

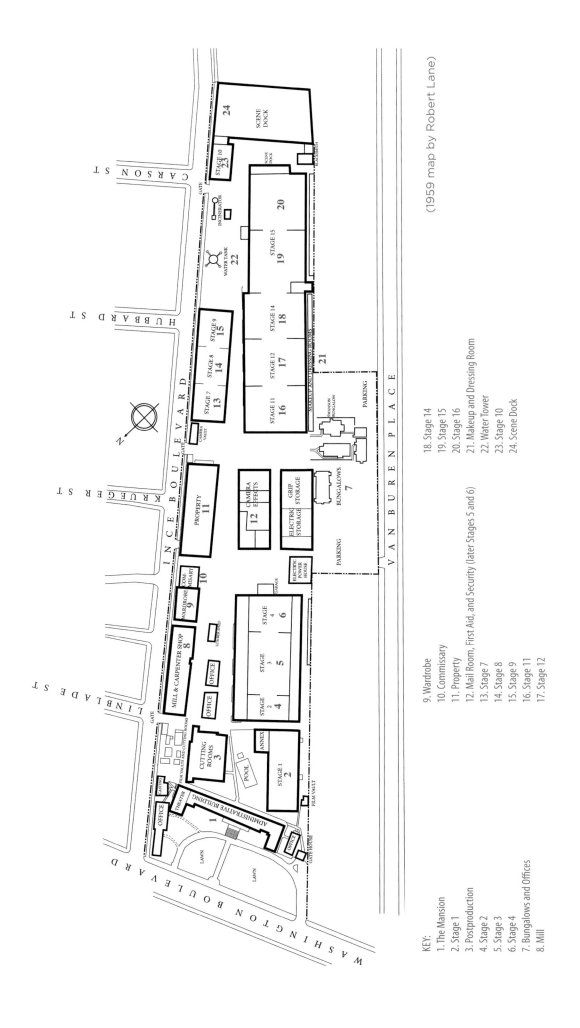

Maps are included, circa late 1950s, so film buffs, adventurous readers, cultural historians, and TV-obsessed insomniacs can now follow along and marvel as the myriad and tangled sets inside this once-celebrated and yet today largely forgotten film factory roll by in swirling and surreal—and let's admit it, often clichéd—cinematic tableaus.

Welcome.

1. The Mansion

In 1918, as the Mansion was being completed, the *Los Angeles Sunday Times* reported, probably because Thomas Ince almost certainly arranged for them to do so, that:

> Thousands of automobile drivers traveling on Washington Boulevard between Los Angeles and the west beaches have seen and admired a large and beautiful mansion of the southern colonial type rising on the west side of the highway in Culver City, and now nearing completion. Probably only a small percentage of them have known that the building was not the mansion of a fine private estate but the great new photoplay studio of Thomas H. Ince. The plant, in fact, is unique among its kind in point of architectural conception, giving no evidence of its individual character from the front or boulevard side.[2]

As noted in a previous chapter, the building was designed by Ince to be something special. He wanted to signify that the motion picture industry had arrived, that the movies were no longer the hardscrabble industry it had been when *he* had started out. Although, truthfully, the building itself still has the vague whiff of a confidence trick hanging over it. Impressive from Washington Boulevard as it is, and gorgeously appointed inside, the interior is actually almost comically shallow. At its narrowest, the mansion is hardly more than 16 feet deep, suggestive more of a backlot than "the White House of silent Drama," as the Ince newsletter often referred to it.

The Mansion, or administration building (now known as Building C), fronting 9336 Washington Boulevard in Culver City, is a two-and-a-half story structure in the "plantation colonial" style, the design of

Every owner of the studio has featured the Mansion prominently in their publicity. This 1923 issue of *Silver Sheet*, a publicity news magazine published by Thomas Ince, is an early example.

The staircase inside the Mansion in the 1920s, and in 2017. A portrait of George Washington was once mounted on this wall.

which, again as noted earlier, seems to have been influenced by George Washington's Mount Vernon. The building was not, of course, Tara from *Gone with the Wind*, in spite of what many have claimed. The structure did appear twice in that movie, however, as the Selznick International logo and again as the house Rhett builds for Scarlett in Atlanta, though in the latter it was disguised by a matte painting, and only the driveway is actually visible on-screen. A better look at the Mansion is available in *Captains Courageous* (MGM 1937), and at the opening of every Selznick International picture.

That driveway also figured in the dramatic preproduction process for *Gone with the Wind*, when every actress in the country, and a great many non-professionals as well, tried to claw their way up the driveway and through the Mansion's doors to audition for the coveted role of Scarlett O'Hara. Someone even put a sign up on the driveway: "Scarlett Way." When the film finally went into principal photography, on January 26, 1939, a Confederate flag was hoisted up the flagpole on the lawn. The same exact flagpole, as well as the lawn and the Mansion, are still there; the sign and the flag are not.

Ince would certainly have been pleased to know that the use of the Mansion's driveway in *Gone with the Wind* was followed by an appearance in the *other* most iconic American film of all time. For *Citizen Kane* (RKO 1941), the White House lawn was in fact played by his own administration building's front yard.

The offices on the northern side of the administration building were built by DeMille in 1926–27 in the same stately colonial style. They were modified extensively by Selznick in 1935, and are often now referred to as both Building D and as the Selznick Wing.

The rarely photographed back side of the Mansion, seen here in 1924 and 2013, originally contained dressing rooms.

In addition to offices, the Mansion and annex includes two screening rooms, the DeMille Theater and the Ince Theater. The Ince Theater is supposedly haunted. Late at night security guards have reported seeing the ghostly apparition of none other than Thomas Ince himself, walking into the projection booth on the second floor. It is indeed fortunate that twenty-first-century studio security guards are able to recognize the builder of the studio, or his ghost, by sight.

The back of the Mansion originally consisted of a long breezeway dotted with cubicle-size dressing rooms, all of which have since been replaced with a more thematically consistent colonial look. Behind the Mansion, at the time of our survey, there was a swimming pool that dated to 1919. The pool area looked like a recreational spot, and it certainly fit well with Ince's newly genteel ideals about the industry. But the structure, make no mistake, was a production tool and was used to store water as an industrial reservoir for safety purposes. Underground pipes ran to the Mansion and out to the soundstages from the basin as a fire preventative, and for production purposes. As a set, the pool appeared in movies like *Hail the Woman* (Associated Producers Inc. 1921), as a setting for a garden party scene, and later in *The Blue Danube* (Pathé 1928), *Suicide Fleet* (RKO 1931), *Nothing Sacred* (United Artists 1937), *The Young in Heart* (United Artists 1938), *Honeymoon* (RKO 1947), and even on television as late as a 1962 *Untouchables* TV episode, "Flowers for My Love."

The basin was 6 feet deep, but if the oldest of studio maps is to be believed, its shape varied somewhat in the early years, when it was slightly larger and more irregular than the more-rectangular reservoir it would become during the RKO and Selznick eras, which makes one wonder if there were two or more pools built consecutively in the same spot. The pool was often covered over when it was not being used, and in fact had been covered and almost forgotten about when it was rediscovered, and removed, in 1988 for producer Grant Tinker's underground parking structure.

Behind where the pool would have stood in the early 1960s could be found the first glimpse afforded a visitor of the studio lot looking east. The view, and many views yet to be discovered while canvassing the lot, will seem familiar to even those

The studio pool, actually a fire protection reservoir, is seen here being filled by Thomas Ince himself. (1922)

Stage 1, seen behind and to the right of the Mansion, as it looked in 1926.

who have never before been through the gates, because these studio streets have, of course, been used as studio streets numerous times. *A Star is Born* (United Artists 1937), Selznick's bittersweet valentine to Hollywood, sadly showed us little of the lot, but other projects such as *What Price Hollywood?* (RKO 1932), *Won Ton Ton: The Dog Who Saved Hollywood* (Paramount 1976), *Moviola: The Scarlett O'Hara War* (TV 1980), *Under the Rainbow* (Orion 1981), *Three Amigos!* (Orion 1986), *Sunset* (TriStar 1988), *Winter People* (Columbia 1992), *Lady Boss* (TV 1992), *Wag the Dog* (New Line 1997), *Action* (TV 1999–2000), *Sunset Action* (TV 1999–2000), *The Lot* (TV 1999–2001), and *For Your Consideration* (Warner Bros. 2006) cast the studio as a studio. *Three Amigos!*, for example, gave us a never-again glimpse at the original glass Stage 1, only two years before its demolition.

2. Stage 1 (70 × 180 feet, 12,600 square feet)

This original stage was built, in 1919, with largely glass walls and ceilings in order to take advantage of natural light. Because of the cantilevered angle it had originally been placed at behind the Mansion, the building was moved into a slightly different position nearer to the property line in 1926 for construction of Stages 2, 3, and 4.

Stage 1's interior as it looked in 1971.

This stage and Stage 2 both offered a unique series of canvas-covered movable flaps, balanced by weights, which could regulate the amount of light from outside being admitted to the set. By raising and lowering these panels, sunlight could be easily controlled, blocked, or diffused at the filmmakers' will.

After talkies arrived and using natural light was no longer practical, Stage 1 was used mostly for costume and equipment storage and was further divided into editing suites and (upstairs) offices, although some of the miniatures for *Rebecca* (United Artists 1940), for example, were actually shot on the stage. Even into the 1980s, the building was still respectfully listed on studio maps as "Stage 1"—although it was no longer used as such.

On Stage 1's northeastern wall, there long stood a smaller building, sometimes known as the "Director's Office," which jutted out like a blister from the main structure. This annex was constructed as a scene dock, but because of its size, it quickly was divided into offices and eventually

Stage 1 in 1979. The annex used for storage and special effects is seen on the left.

became one of many homes for the grip department. Returning to the lot in 2009 for a special visit to talk to local children for Sony Animation, special effects maestro Ray Harryhausen recalled that many of the Oscar-winning animation composites for *Mighty Joe Young* (RKO 1949) were, in fact, crafted over fourteen months in that little room on the side of this stage.

Perhaps because of its vintage, its longevity, and its unusual antiquated look, persistent ghost stories have long been whispered on the lot that seem to involve Stage 1. For example, when the building was finally scheduled to be demolished (which would happen in March of 1988), Eugene Hilchey, a mechanical effects technician who had earlier played a small part in the saga of the last days of 40 Acres, found himself on the lot and decided to take some photographs of the soon-to-be-destroyed building. Hilchey was poking about with his camera under the amber sunlight streaming in through the old glass walls when he realized that suddenly the room had become very cold—even though it was 80 degrees outside. Unnerved, he went outside to his car to retrieve his jacket. When he returned a cold gust of wind was blowing across the stage, even though the doors were closed. He could also hear footsteps on the catwalks above him in the rafters, even though he knew he was the only person in the building. Hilchey did not linger long!

Stage 1 is demolished. (1988)

Shortly thereafter, even as Stage 1 was literally being bulldozed into oblivion, studio employees could be seen scurrying out from the other side of the building, carrying whatever odd artifacts had not been removed before the structure had been red-tagged. John Bertram, a studio historian and stage manager, was later shown a detailed *USS Maine* portrait, probably painted in the late 19th century—which management did not feel was important enough to salvage—which had been carried away by an employee who was there.

Today there is no Stage 1. A parking lot and some production support buildings stand on the site.

3. Postproduction

This large, oddly imposing structure was built, circa 1920, as the most modern and complete film laboratory on a studio lot in the world. It also included a fireproof film

STAGES 2, 3, AND 4

The *original* Stage 2, built in 1918, was, like Stage 1, a glass stage 72 by 180 feet across. Unlike Stage 1, which would haunt the lot for generations from basically its original location, this Stage 2 would be moved across the property to the center of the lot, circa 1925, so that the current Stages 2, 3, and 4, all in one enormous building, could go in. This original stage, later known as Building P, would be demolished during the Grant Tinker makeover, when Stages 5 and 6 would be added.

The original Stage 2 and its annex are moved to the center of the lot to create Stages 2, 3, and 4. (1925)

This Stage 2, along with its sister Stages 3 and 4, would be built in 1925, presumably for *The King of Kings*. Conveniently, on-lot office space was here, supplemented by building offices along the face of the stage with second-floor cubicles above them.

This building would be the first stage on the lot to be converted for sound recording, an expensive process that involved opening up the walls and padding them with thick buffering materials in order to keep the outside world's caterwauling outside. Accommodations also had to be made for the serpentine cables that the sound captured on the set was recorded into, which initially had to be piped outside to a separate bunker filled with confused technicians.

Like many buildings on the lot, Stages 2, 3, and 4 are supposedly haunted. In this case the ghost is a man in a bowler hat who

Stages 2, 3, and 4 as they appeared in 1933.

52 HOLLYWOOD'S LOST BACKLOT

The imposing western wall of Stages 2, 3, and 4 is visible from far outside the lot, and has often been used as a place for the studio's name, whatever that name may have been at the time. (1988)

walks through walls and was once supposed to have exclaimed, "I don't like what you are doing to my studio."

There is a dividing wall with a 44-foot-long door between Stages 2 and 3 so that large sets for large projects can creep over into the adjacent stage(s), although, for reasons no one has been able to explain, the floor of Stage 2 is 3 to 8 inches higher than that of Stage 3. In 2003, for the series *Las Vegas* (TV 2003–8), the wall between Stages 3 and 4 was removed entirely to create one gigantic 219 by 119 foot (26,061 square feet) shooting space, called simply Stage 3. Scheduled for 2019 as part of the ongoing redevelopment of the studio is the replacement of Stages 2, 3, and 4 with a proposed 170,800-square-foot "Digital Media Building."

The best way to explore Stages 2, 3, and 4, and the rest of the studio, is through the projects that were crafted there. For the most iconic or well-traveled productions, a separate listing, for each title, for each location, is provided. For other tenants, that title is only listed once, although other stages that film wandered into are also noted, whenever possible, under that film's title. Clarification as to how a film used a space has been provided, again whenever possible.

Let's go inside.

4. Stage 2 (55 × 120 feet, 40 feet high, 6,600 square feet)

Campus Champs (RKO 1931): short subject, shot here and on 40 Acres.

Disappearing Enemies (RKO 1931): short subject, shot here and on Stage 1.

Chasing Trouble (RKO 1931): short subject, shot here.

Twisted Tales (RKO 1931): short subject, shot here.

Devotion (RKO 1931): Leslie Howard, in perhaps his first of many roles at the studio. The film was also shot on Stages 3, 4, and 11.

All Gummed Up (RKO 1932): short subject, shot here.

Stage Struck (RKO 1932): short subject, shot here and on Stage 4.

Parents Wanted (RKO 1932): short subject, shot here and on Stage 4 and 40 Acres.

Sin Takes a Holiday (Pathé 1932): The ocean liner and drawing room scenes were shot here. The Parisian nightclub and drawing room scenes were shot on Stage 3, the New York hotel scenes on Stage 4, and "Sylvia's NY apartment" on Stage 9.

Strictly Dynamite (RKO 1934): Jimmy Durante in a radio "broadcast station."

Gone with the Wind (MGM 1939): Leslie Howard returns. Shot here were assorted Tara interiors, and also Melanie's living room. As could be imagined, this one shot all over the lot, as noted below.

Intermezzo: A Love Story (United Artists 1939): Leslie Howard again. And an effervescent young Ingrid Bergman made her American film debut in this remake of an earlier (1936) Swedish picture, which she had also starred in. Tests for the "Int. Library" scenes were shot here. Later the cabin scenes aboard the liner were recorded here as well. As noted below, the movie shot in other stages too.

Fantasia (RKO 1940): In January 1938 Disney Studios rented the stage for "rehearsing, scoring and/or recording music" with Leopold Stokowski for the legendary "Sorcerer's Apprentice" number for what would become one of the most experimental films ever released by a Hollywood studio. Live-action footage with Stokowski was also shot here, although it was later replaced with footage shot at the then (1940) new Disney Studios of Mickey Mouse shaking hands with the maestro.

Rebecca (United Artists 1940): The location for the "gate and drive," the "English mortuary," and part of the dance are all here. The film shot all around the lot, as noted below.

David O. Selznick consults with Alfred Hitchcock on the bedroom set of *Rebecca*. (1940)

Father Takes a Wife (RKO 1941): assorted scenes shot here and on Stage 3.

A Date with the Falcon (RKO 1942): theater set scenes shot here and on Stage 3 under the title "The Falcon Takes a Wife" in 1941.

Petticoat Larceny (RKO 1942): assorted scenes.

Radio Runaround (RKO 1943): assorted scenes for short subject.

Around the World (RKO 1943): Popular-at-the-time bandleader Kay Kyser entertains troops across the globe, without actually leaving RKO. On this stage the "clipper" scenes were shot. Other sequences were shot on Stages 3, 4, and 15.

Action in Arabia (RKO 1944): also shot on Stage 15 and in the studio paint shop.

Banjo (RKO 1946): "Ext. Warren kitchen," also shot on Stage 8 ("Pat's bedroom") and 40 Acres.

American Creed (1946): This late-on-the-scene propaganda film, produced by Selznick under the shooting title "Brotherhood," used an all-star cast (James Stewart, Van Johnson, Ingrid Bergman, Jennifer Jones, Katharine Hepburn, and even Shirley Temple, among still others) to preach tolerance among religions. The film also shot some scenes on the ever-popular Mansion lawn.

The Farmer's Daughter (RKO 1947): Scenes were shot here and on Stages 3, 4, 7, 8, 9, 11, and 14 and at MGM and the Columbia Ranch. David O. Selznick had intended to produce as a vehicle for Ingrid Bergman, before selling the project to RKO. Ultimately Loretta Young collected an Oscar for her work.

Portrait of Jennie (SRO 1948): Assorted interiors were shot here, including the Coast Guard headquarters, Gus's apartment, the bookstore, and the marine hardware shop. Other stages where this film shot are detailed below.

The Velvet Touch (RKO 1948): Theater and theater backstage scenes were shot here and on Stage 3. Other sequences were shot on Stages 4, 7, and 8.

Race Street (RKO 1948): assorted scenes.

Gambling House (RKO 1950): "backstage and dressing room" shot here; scenes also shot on Stage 4 and 40 Acres.

Walk Softly, Stranger (RKO 1950): assorted scenes shot here and on Stage 12.

My Forbidden Past (RKO 1951): The "inquest room corridor" was here, but the film shot all over the lot, for fourteen days, on Stages 3, 7, 9, 11, and 12.

Two Tickets to Broadway (RKO 1951): Combined with Stage 3, this was the interior of a TV studio.

Lassie (TV 1958–64): The long-running series apparently shot, intermittently, on this stage during these years.

The Untouchables (TV 1959–63): a regular stage for this one, which also regularly used Stages 3, 9, and 14.

Batman (TV 1966–68): occasionally shot here; also used other stages as detailed below.

Motel Hell (United Artists 1980): scenes shot here and on Stage 7.

Palmerstown, U.S.A. (TV 1980–81): Best-selling author Alex Haley created this sincere, Depression-era drama, which shot both here and next door on Stage 3.

Strangers Kiss (Orion 1983): Blaine Novak, the writer of this low-budget movie about the making of a low-budget movie, was interviewed in the *Los Angeles Times* during its production. The interview described the (then) Laird Studio as "homey, slightly ramshackle" and Novak as saying that "when you're making a movie you feel you can break the rules of real life. You feel protected from the outside world."[3] These sentiments are understandable because for this project then-owner Joseph R. Laird quite generously gave the production the use of his facilities, including this stage and Stage 3, for nothing up front.

Mass Appeal (Universal 1984): This Jack Lemmon vehicle shot here and on Stage 3.

All of Me (Universal 1984): Steve Martin and Lily Tomlin vehicle.

Radioactive Dreams (De Laurentiis Entertainment Group 1985): shot here and on Stages 3 and 4.

Prizzi's Honor (20th Century Fox 1985): assorted scenes shot here and on Stages 3 and 8.

Northstar (TV 1985): failed pilot, shot here and on Stage 4.

Back to School (Orion 1986): assorted scenes shot here and on Stages 3, 11, and 12.

Child's Play (United Artists 1988): The first of the long-running "Chucky" killer doll series shot here and on Stages 3 and 4.

Baywatch (TV 1989–99): shot here and on Stages 3 and 4 for the first season, at the end of which the show's interiors were moved to a (cheaper) warehouse in Marina del Rey.

The Two Jakes (Paramount 1990): This *Chinatown* (Paramount 1974) follow-up shot here and on Stages 10, 14, and 16.

Dracula: Dead and Loving It (Columbia 1995): "Borgo Pass" scenes were shot here and on Stages 3 and 4 for Mel Brooks's visit to Transylvania.

Flubber (Disney 1997): one day of stunt-rig shots only.

Fallen (Warner Bros. 1998): assorted scenes shot here and on Stages 3, 8, and 9.

Teaching Mrs. Tingle (Miramax 1999): assorted scenes shot here and on Stage 3.

Las Vegas (TV 2003–8): built a full-size Nevada casino here and in Stages 3 and 4.

The Last Ship (TV 2014–): also used Stages 3 and 4 for single sets.

5. Stage 3 (120 ×140 feet, 35 feet high, 16,800 square feet)

Officer O'Brien (Pathé 1930): assorted scenes shot here and on Stages 2, 8, 10, 11, and 14.

Bad Company (RKO 1931): shot all over the lot and on 40 Acres.

Stage 2 being utilized for the TV series *The Last Ship*. (2015)

The Tip-Off (RKO 1931): assorted scenes shot here and on Stages 4, 8, 11, and 12.

Open House (RKO 1931): short subject, also shot on Stage 7 and 40 Acres.

Flaming Gold (RKO 1932): "Int. birdcage" was here; also shot scenes on Stages 4 and 9.

Bird of Paradise (RKO 1932): "Int. yacht, cabin" was here.

Born to Love (RKO 1932): assorted scenes shot here and on Stages 1, 9, and 10 and 40 Acres.

The Most Dangerous Game (RKO 1932): The famous, and censored, scenes of a trophy room filled with mounted human heads were shot on this stage, as were most of the interiors of evil Count Zaroff's castle. Other scenes were shot in other stages, as noted below.

Bed of Roses (RKO 1932): "Jackson Square sequence" shot here; "Warden's office" was on Stage 11.

Morning Glory (RKO 1933): "Int. Easton Theater—audience & backstage."

Topaze (RKO 1933): assorted scenes.

The Adventures of Tom Sawyer (United Artists 1938): The memorable graveyard murder scene was shot here. Other locations across the lot are detailed below.

Bringing Up Baby (RKO 1938): On December 20, 1937, and encircling that Christmas, namely the 23rd, 24th, and 29th, the Howard Hawks screwball classic shot the Ritz Plaza Hotel lounge and bar scenes on this stage and on Stage 4.

Made for Each Other (United Artists 1939): Inserts and miniatures, as well as the Catholic hospital, Central Park, "Int. Apartment," and process shots of the terrain as seen from a plane cockpit, were realized here. The film also shot on other stages, as noted below.

Intermezzo: A Love Story (United Artists 1939): The Swedish concert hall, including the boxes and balcony, was here. Along with Stage 4, this stage was later used for "embankment" scenes.

Rebecca (United Artists 1940): Assorted Manderley interiors were shot here and on Stage 4.

Triple Justice (RKO 1940): The "jail," the "shack," and the "town hall" were here. Stage 9 hosted the "Wells Fargo office" and the "saloon back room."

Tom, Dick and Harry (RKO 1941): The motion picture theater scenes were shot here. The film also shot on Stage 8.

Let's Make Music (RKO 1941): assorted scenes.

Citizen Kane (RKO 1941): The 1919-set opera house scenes were shot here. Other stages *Kane* made history upon are detailed below.

Footlight Fever (RKO 1941): The "theater" and "Avery's dressing room" were here.

The Magnificent Ambersons (RKO 1942): The ballroom, Isabel's room, the club room, the "second hall," the barbershop, "George's room," "Bronson's room," and "Jack's bath" were built and shot here for Orson Welles's troubled, brilliant *Citizen Kane* follow-up. Other stages utilized in the production over the course of its fifty-two days on the lot follow below.

Here We Go Again (RKO 1942): The "poolroom" and the "silvertip lobby" were here. "Upington Hall" was on Stage 12.

Seven Miles from Alcatraz (RKO 1942): also shot on Stages 2 and 15 and in the studio Mill.

Mexican Spitfire's Blessed Event (RKO 1943): Lupe Vélez made seven "Mexican Spitfire" movies for RKO, of which this was the last. Interiors were shot here, and "desert" exteriors on Stage 4. Tragically, Vélez would commit suicide the following year.

Since You Went Away (United Artists 1944): Most of the Hilton family home interiors for this well-made glorification of the WWII home front were here and on Stage 4 and other stages, as noted below.

Bride by Mistake (RKO 1944): assorted scenes shot here and on Stage 14.

A Game of Death (RKO 1945): Rainsford's room, Ellen's room, and the trophy room were all constructed here for this remake of *The Most Dangerous Game* (RKO 1932)—shot on the same stage. The "dining salon" was on Stage 7, and some jungle scenes were shot on Stage 14.

First Yank into Tokyo (RKO 1945): Tom Neal played the yank in Tokyo for the "edge of the woods" and "Bataan shrine" scenes created here. Other sequences were shot on Stages 2, 4, 7-8, 9, 12, and 14 and on 40 Acres across twenty busy days on the lot.

They Won't Believe Me (RKO 1947): with Stage 4, for "falls and pool."

Night Song (RKO 1947): Process shots were composited here; other scenes were shot on Stage 4.

Fighting Father Dunne (RKO 1948): Pat O'Brien, naturally, played the title role. Stage 4 played the "theater balcony" and "O'Donnell's home."

Portrait of Jennie (SRO 1948): Inserts of paintings in progress by artist Joseph Cotton were here. His studio was also here, as was "Spinney's office."

If You Knew Susie (RKO 1948): comic Eddie Cantor, in a late-career curtain call; also Stages 2 and 3.

The Judge Steps Out (RKO 1948): shot here and on Stages 9 and 15 in 1947 under the title "Indian Summer."

Armored Car Robbery (RKO 1950): assorted scenes.

Where Danger Lives (RKO 1950): assorted scenes; also shot on Stage 4 and 40 Acres.

The Greatest Show on Earth (TV 1963–64): shot here and on Stage 4.

Blood Beach (The Jerry Gross Organization 1980): assorted scenes; also shot on Stage 2.

The Jayne Mansfield Story (TV 1980): '70s sex symbol Loni Anderson and then–body builder Arnold Schwarzenegger are certainly well cast as '50s sex symbol Jayne Mansfield and then–body builder Mickey Hargitay.

Raging Bull (United Artists 1980): boxing scenes only.

Dead Men Don't Wear Plaid (Universal 1982): Assorted and familiar film noir tropes were variously ransacked, re-created, and parodied here, often using the actual footage from dozens of 1940s melodramas—some of which had, perhaps intentionally, perhaps not, been shot on this very same stage—with Steve Marin edited into the action on cleverly re-created sets, here and on Stage 12.

E.T. the Extra-Terrestrial (Universal 1982): Most of the family home (and note that the family's name was unmentioned anywhere in the film) was constructed on this stage, with additional sets on Stages 2 and 4, although what today is arguably Steven Spielberg's archetypical film was not originally scheduled to film in Hollywood at all. Production manager Wallace Worsley Jr., whose father had directed the Lon Chaney *Hunchback of Notre Dame* (Universal 1923)—which, come to think of it, bears some similarities to *E.T.*—had originally scheduled this little film about a lonely boy and a lonely alien to shoot at the Osmond Studios in Orem, Utah! Assistant director Charles Ziarko, however, convinced Spielberg that shooting locally, at the Culver Studios, could save the production as much as a million dollars in location, hotel, and per diem expenses—money which was eventually repurposed for a location trip to Redwood National Park in Northern California.

E.T. apparently reads the trades. (1982)

The Man with Two Brains (Warner Bros. 1983): Steve Martin and director Carl Reiner (*The Jerk* and *Dead Men Don't Wear Plaid*) return to the lot for a third, and very funny, time. This one was also shot on Stage 4.

A Bunny's Tale (TV 1985): Kirstie Alley as Gloria Steinem!

Short Circuit (TriStar 1986): assorted scenes shot here and on Stages 2 and 11.

House II: The Second Story (New World 1987): assorted scenes.

Allan Quatermain and the Lost City of Gold (Cannon 1986): assorted scenes, also Stage 11.

TV 101 (TV 1988–89): assorted scenes.

Judgment Night (Universal 1993): assorted scenes, also Stages 2 and 4.

North (Columbia 1994): The lightning strike on a tree was shot here.

Town and Country (TV 1996): Assorted scenes were shot here for this series that's unrelated to the 2001 feature film (except by title), which would later shoot on the same stage.

Sour Grapes (Columbia 1997): assorted scenes, here and on Stage 4.

Wag the Dog (New Line 1997): assorted scenes, here and on Stage 5.

Mad City (Warner Bros. 1997): assorted scenes.

Starship Troopers (TriStar 1997): "space shuttle."

Krippendorf's Tribe (Disney 1998): assorted scenes, here and on Stage 2.

Godzilla (Columbia 1998): shot unknown scenes for one week on stage.

Town & Country (New Line Cinema 2001): Warren Beatty's expensive, disastrous comedy train-wreck shot most of the New York fashion show sequences here, among other scenes shot here and on two other stages (11 and 14). The project would ultimately stagger along on the lot, and on other lots, for months of endless reshoots, reimaginings, and rewrites. Ultimately, before the film would finally be completed, months behind schedule and fortunes over budget, *Town & Country* would ensnare many, many expensive stars. In addition to ringleader Beatty, the unwieldy, though admittedly most interesting, cast eventually boasted of Diane Keaton, Nastassja Kinski, Goldie Hawn, Jenna Elfman, Charlton Heston, Gary Shandling, Andie MacDowell, Buck Henry, Holland Taylor, and even a young Josh Hartnett. Reportedly, their lavish production trailers eventually filled up the entire lot, and ultimately spilled outside onto Ince Boulevard.

Stages 2, 3, and 4 as they looked in 2015.

Philly (TV 2001–2): The Kim Delaney drama was shot here.

The Polar Express (Warner Bros. 2003): The then-experimental, ground-breaking motion-capture scenes were recorded here.

Extant (TV 2014–15): Assorted scenes were shot here, combined with Stage 4, for this Halle Berry starrer. Steven Spielberg was an executive producer.

6. Stage 4 (75 × 120 feet, 35 feet high, 9,000 square feet)

Rebound (Pathé 1931): The Truesdale dining room, stateroom, bedroom, and bathroom—in fact, most sets for this one—were here, although remarkably the film also made appearances on Stages 2, 3, 4, 7, 8, 9, and even, for retakes, the seldom-used Stage 10.

The Messenger Boy (RKO 1931): short subject, also shot on Stage 7 and exteriors on 40 Acres.

Lonely Wives (RKO 1931): also shot on Stages 3, 6, 7, and 10.

Beach Pajamas (RKO 1931): short subject, also Stages 7 and 9.

Hold 'Em Jail (RKO 1932): Comedy team Wheeler and Woolsey used the stage as "Maloney's office." Stages 7 and 8 were also utilized.

Peck's Bad Boy (RKO 1934): assorted scenes.

The Adventures of Tom Sawyer (United Artists 1938): "courtroom."

Intermezzo: A Love Story (United Artists 1939): "Swedish café" and "Holger's room"; also, inserts of Leslie Howard's picture on the wall were grabbed here.

Made for Each Other (United Artists 1939): some miniatures, as well as the full-size casino set.

Gone with the Wind (MGM 1939): Many smaller and less-used sets were constructed here for Selznick's super-production—for example, Aunt Pittypat's parlor, where Rhett proposes to Scarlett; the hallway and internal doors of Tara; and the "sleepytime" sequence at Twelve Oaks. Also, the tiny London hotel room sequence, the only part of the film not set in the American South, was shot here, with Big Ben (actually a 9 1/2-foot-tall cutout) rather too obviously visible through the bedroom window.

Rebecca (United Artists 1940): Manderley lower floor interiors were constructed here. Inserts of the "invitation to the ball" and Maxim's address book pages were also shot here later by a second unit.

My Favorite Wife (RKO 1940): The courtroom scenes were shot here, with other sequences on Stages 12 and 14.

Lucky Partners (RKO 1940): Assorted scenes shot here and on Stages 12 and 14.

Beyond Victory (RKO 1941): Also shot on Stages 6, 8, and 9 and on 40 Acres.

A Girl, a Guy and a Gob (RKO 1941): Lucille Ball, who would one day own the studio, costars.

Seven Days Leave (RKO 1942): assorted scenes.

This Land is Mine (RKO 1943): "von Keller's headquarters." Von Keller is played by Walter Slezak in this French-set Maureen O'Hara vehicle.

Spellbound (United Artists 1945): The casino dream scenes, designed by Salvador Dali, were shot here. Other scenes were shot across the lot, as noted below.

Nocturne (RKO 1946): "detective offices," "police headquarters," and "Warne's home" sets for this George Raft film noir, which also shot on Stage 15.

The Locket (RKO 1946): assorted scenes; also shot on Stages 3 and 15.

Dick Tracy Meets Gruesome (RKO 1947): Boris Karloff, naturally, plays Gruesome.

Crossfire (RKO 1947): "theater building" and "small joint and yard."

Good Sam (RKO 1947): also shot on Stages 9, 11, and 12.

Magic Town (RKO 1947): also shot on Stages 3, 7, and 14.

Seven Keys to Baldpate (RKO 1947): Assorted scenes for this oft-filmed chestnut were recorded here and on Stage 3.

Hunt the Man Down (RKO 1950): three days only.

Never a Dull Moment (RKO 1950): assorted scenes, here and on Stage 3.

The Midnight Hour (TV 1985): Halloween special.

Blind Date (TriStar 1987): assorted scenes, with Stages 7 ("art gallery"), 9 ("disco"), and 11.

Body of Evidence (TV 1988): assorted scenes, with Stage 11. Not to be confused with the kinky 1993 feature film (with Madonna), which would shoot here as well.

Marilyn & Bobby: Her Final Affair (TV 1993): Melody Anderson played Marilyn (Monroe) and James F. Kelly was Bobby (Kennedy), a role he specialized in. Kelly would later play JFK as well, in *Sinatra* (TV 1992).

Ladies Man (TV 1999–2001): Alfred Molina had the title role.

City of Angels (TV 2003): The hospital sets for this medical drama were primarily here, although the show also shot on interconnected sets on Stages 2, 3, and 4.

Baskin-Robbins commercial (1995): identified in the production records as being titled "Crossroads."

Air Force One (Columbia 1997): assorted scenes.

7. Bungalows and Offices

The studio bungalows are freestanding buildings originally used as dressing rooms and now utilized as boutique office space. As they stand all over the lot, as noted on the map, they will also be designated here with their studio building numbers or letters (different regimes have used both) for reference.

Many famous and forgotten, and noted and notorious, people have occupied various spaces on the lot for many years. Some rented the bungalows, some were in the Mansion, some occupied offices across the lot. In many cases it is no longer known or cannot be verified where these companies or individuals were located. What follows is a partial list, a roughly chronological sampling of some of the companies and people that have occupied these spaces. It should be noted that in some cases the projects they produced were shot at the studio, while in other cases they were not.

In 1932 it was noted in studio records that Oliver Morosco Films was on the lot. Morosco (1875–1945) was a fascinating Broadway impresario who produced some early films, the last of which would be in 1919, so it is interesting to speculate what he was doing on this lot at this late date, and which productions he was never able to get off the ground in Culver City.

At the same time, Edward R. and Victor Halperin, two brothers, were also on the lot. Today, the Halperins are best known for the fascinating cult horror favorite *White Zombie* (United Artists 1932), which was shot at Universal, after which they slipped into relative obscurity.

Edward Small's Reliance Pictures was also on the lot at the time. Small, like the Halperins, had a distribution deal with United Artists. In 1936 he moved to RKO and would eventually became an independent producer and powerful agent.

In March of 1946 Small was still on the lot, as was RKO; Selznick; Paramount cofounder Jesse Lasky's production company, Rainbow Productions, which had just produced *The Bells of St. Mary's* (RKO 1945); and something called Liberty Productions, which might have referred to Frank Capra's Liberty Films, which had just produced *It's a Wonderful Life* (RKO 1946).

In 1948 Argosy Pictures, which belonged to director John Ford and producer Merian C. Cooper, moved onto the lot. (Cooper had previously, in 1933–34, succeeded Selznick as production head of RKO, and then had become a second-floor tenant of the Mansion as a vice president of Selznick International.)

In 1955 Four Star Productions was a tenant. The four stars were, at the time, David Niven, Dick Powell, Charles Boyer, and Joel McCrea, although the company ended up making mostly television Westerns.

The tenants continued to parade through. In 1958 NTA (National Telefilm Associates) was renting space on the lot. NTA was a television syndication company that occasionally coproduced original syndicated content as well. Some of this product, like *Whirlybirds* (TV 1957–60), *Official Detective* (TV 1957–58), and *This Is Alice* (TV 1958), was shot on the lot. The same year, Rolland D. Reed Productions, which rented editing equipment, leased space on the property. In 1962 George Stevens, shooting his gargantuan *Greatest Story Ever Told* (United Artists 1965) on the lot, rented most of the office space at the studio as well.

Of course, some of the occupants of these offices were considerably more obscure than John Ford and George Stevens. In 1968 National General Corporation rented space behind the gates from Paramount, the landlord of the moment. Haley Tat Productions put up its banner in 1979. Tom Laughlin, the renegade filmmaker who scored an international counterculture hit with *Billy Jack* (Warner Bros. 1971), moved onto the lot in the mid-1970s, although only *The Master Gunfighter* (Taylor-Laughlin Productions 1975) actually filmed on the stages there. Mel Brooks leased what is now known as Building B, which is in front and south of the Mansion, during this same chaotic era—and is still there.

During the same era, a company called Airline Film & TV Promotions, a subsidiary of aviation giant TWA, was based on the property and started to rent airline fuselages out to the film industry as sets. Originally the actual cabins and cockpits were stored in warehouses and the client was responsible for securing a production location, but the company was so successful that the studio's Stage 16 was more or less permanently set aside for aviation-themed productions during this era.

In 1984 the production companies on the lot included H.K.M. Productions, Fred Levinson Productions, Big City Productions, Image Steam Productions, and Vern Gillian Productions—all of which were primarily producing

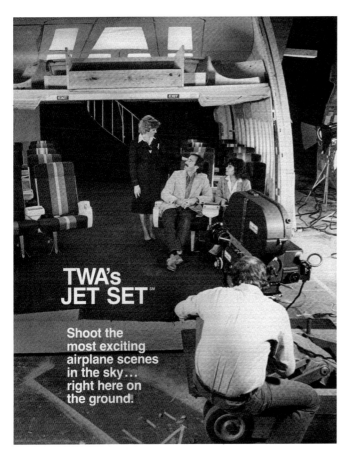

In the 1970s and '80s TWA rented office and soundstage space on the lot, utilizing training fuselages as interiors for movies such as *Airplane*. (1980)

television commercials. Yet at the same time, Warner Bros., Norman Jewison, Rob Reiner, and Sylvester Stallone were jostling about in adjacent offices as well.

In 1987 Blake Edwards was renting several offices on the lot and directing *Sunset* (TriStar 1988) nearby. Geffen Films was renting office and soundstage space for Tim Burton's *Beetlejuice* (Warner Bros. 1988). TriStar was mounting *Sweet Hearts Dance* (1988), and Tobor Pictures (spell it backwards) was completing *RoboCop* (Orion 1987). Grant Tinker, who moved onto the lot that year, displaced producer Bud Yorkin when Tinker decided he wanted Yorkin's office.

Two of the bungalows, Buildings H and I, are among the most historic quarters on the property, although neither of them is in its original location. They were originally next to each other, behind the Mansion. "H" is the older of the two, going back, apparently, to the Ince era. "I" was built during DeMille's tenancy. Both buildings were salvaged and relocated during the Grant Tinker period when, in 1987–88, extensive renovations to the studio were made, including building the underground 420-car parking lot and adding to the awnings and breezeway on the back of the Mansion, both of which look authentic but largely aren't. Building H supposedly is haunted, with slamming doors, breathing noises, and sighs having been reported by cleaning crews alone in the office at night.

Building S (then behind Stage 11) was allegedly built as Vivien Leigh's quarters for *Gone with the Wind*. Her costar, Clark Gable, reportedly occupied the other side of the same building at the same time, although there is a photograph of the star on the lot that seemingly shows him using a production trailer as his dressing room. The design for the building is attributed to Lyle Wheeler, the art director of that film. However, the building itself seems to have existed as far back as 1929, so perhaps Wheeler only performed a facelift on an existing structure for the stars.

Vivien Leigh's dressing room is one side of this bungalow (here visited by a smiling starlet in 1939); Clark Gable's was allegedly on the other. On the right, the same bungalow as it looked in 2015.

Joseph R. Laird used the building as his office in the early 1980s, and Blake Edwards Productions is also known to have leased this building.

Next door is Bungalow V, which, again allegedly, was once Gloria Swanson's dressing room—supposedly built for her by the smitten Joseph P. Kennedy in 1928. Some maps in the 1950s, long after Swanson would have been gone, do identify the building as the "Swanson Bungalow." Publicist Russell Birdwell claims to have found bullet holes and surveillance equipment in the building's walls—both apparently the work of a jealous Kennedy. Again, Blake Edwards Productions would utilize this building in the 1980s.

Olivia de Havilland supposedly used Bungalow T, which is near the Van Buren Place studio wall. This particular building also appeared on-camera behind John Wayne in *Flying Leathernecks* (RKO 1951). Next door is Bungalow U, once said to be the office of Orson Welles. In the 1950s, long after Welles would have departed, the building housed a steam room.

On what seems to be an uncharacteristically rainy day, director Norman Foster and Orson Welles enter Welles's bungalow during the production of *Journey into Fear*. The bungalow would not be Welles's for much longer. (1942)

The bungalows behind Stage 11 were all moved to a spot behind the Mansion for a 2018 renovation.

The other clapboard cottages in this area have long been used as dressing rooms and (later) production support offices. One of them, the letter of which has apparently now been forgotten, once had a dressing table with a mirror inside, elaborately stenciled with the initials "CBD"—presumably, hopefully, for Cecil B. DeMille. Bungalow R, to the west, was once used by Alfred Hitchcock, although that director apparently never thought to have his monogram inscribed on a dressing table there.

Undocumented though most of these stories are, one certainly *wants* them to be true.

In 2016 Bungalow "R" was removed, and presumably destroyed, to make way for the Van Buren parking lot, which would also be known as parking structure "R." When that lot was expanded in 2018 (to 1,109 spaces) Bungalows S, T, U, and V were carefully moved to behind the Mansion, where, happily, they remain today, to hopefully inspire creativity in some later-day DeMille.

8. Mill

Next to the gate was a long 10,000-square-foot building, constructed in 1927, which replaced an older Ince-era structure used, like its next-door neighbor, as a costume and dressing room building.

Originally called the Mill and Carpentry Shop, or Building J, this was where every physical object that appeared on camera, from canoes to battleships, was crafted. Tara and Manderley and Mayberry and Skull Island and a hundred, perhaps a hundred thousand, pieces, from sets to set dressing, were assembled here.

A Building J is still there, but although the modern version occupies nearly the same footprint as the original, this new Building J is actually an entirely different structure that contains offices. Employees disdainfully call this new building "Alcatraz," because of the oppressive, repetitive, architectural style it was constructed in.

The assembly of the items that actually appear onstage now happens to the east near or in the old Scene Dock, or south of Stage 5, or off-lot and in much smaller quarters, as the dirty, heavy-industry aspects of production are largely being outsourced and deemphasized in favor of a "campus" atmosphere. "If they could call it a studio and just turn it into offices, they would be happy," John Bertram wryly notes.[4]

The studio Mill in 1972.

FRONT LOT: SETS, SETTINGS, AND SET PIECES 69

The same building from outside the studio on Ince Boulevard. (1974)

The Mill occasionally has been used for other purposes. It has been reported that *Rocky* (United Artists 1976) shot scenes there, but apparently the building was only used for rehearsals and training scenes.

9. Wardrobe

The only brick building on the lot, Building L, was built in 1926 by DeMille for wardrobe storage, although the on-lot film laboratory seems to have also used the structure during the early years. Actually, Ince apparently had plans for a similar building in the same spot, although all he lived to see was a much-less-ambitious lumberyard standing on the site before his death. The current building, which, post-DeMille, RKO, Selznick, and Desilu all continued to use for costumes, was also occasionally combined with the makeup department during these eras.

In 1951, when Selznick was trying to appraise the value of his physical assets, his wardrobe department could only come up with $1,101.50 in assets. It should be noted, however, that this laughably miniscule figure only appraised the physical contents of the building and of Stage 1, which was also being used for costume storage at the time. This meant that the coat hangers and sewing machines and wardrobe racks stored there had a tangible value, but the costumes themselves, heirlooms from a hundred movies, were merely a curiosity and not a reportable asset.

The only specific costume piece mentioned in the inventory was "Cellophane hat box & *GONE WITH THE WIND* hat."—appraised at a value of $12.50!

70   HOLLYWOOD'S LOST BACKLOT

The wardrobe department during the Selznick era. (circa 1937) Inset: A men's wardrobe sign recovered in 1989.

Apparently, the hat was mentioned because of the cellophane hat box. Incidentally, no one knows which *GWTW* hat is being referred to here, but an apparently screen-worn Clark Gable top hat from that movie sold at a Christie's auction in the UK in 2014 for the equivalent of $11,377.32 US dollars.

Walter Plunkett, *Gone with the Wind*'s costume designer, who returned to the studio for a visit in 1976, was, even at the time, astonished at the latter-day value and nostalgia audiences and collectors feel about those costumes. "The only thing I remember is that the laundry bill was more than $10,000. It got awfully dusty out there at the plantation," he chuckled.[5]

In the 1980s during the Laird period, when the studio was only a rental lot, the Commissary, next door, took over the old Costume Building as well, using it for catered events—which could be lucratively billed out to tenants. The two buildings were eventually combined by removing the wall between them, although today that

Walter Plunkett, costume designer for *Gone with the Wind*, seen with Scarlett O'Hara herself, Vivien Leigh, in 1939, and on a return visit to the studio in 1976. The mysterious "Grayson Potchuck Productions" sign is actually set dressing for the (then) recent flop *Won Ton, the Dog Who Saved Hollywood*.

wall is largely intact again. The old costume department side has been divided into offices. In the 1990s the studio security department had their nerve center in these offices, as did the 1992–99 TV series *Mad About You*.

10. Commissary

The building where a thousand stars have eaten a thousand good, or not-so-good, meals began its life as an Ince-era effects stage, utilized for "trick shots" as they were called then.

The original Commissary was conveniently near the star bungalows on the southern end of the property. The Commissary (Building M) that stands today was built in 1930–31 and supplemented by an outdoor dining area by Selznick in 1945. As can be imagined, it was often the social center of the studio. Stars could be seen daily dining inside or at tables in front of the building; on-set romances and partnerships flared up and burned out at those tables. Crews and screenwriters and

The Commissary, where thousands of good and indifferent and even downright awful meals have been prepared, is on the right, under the green and white awnings. (1991)

hairstylists and executives and executive mistresses all came together and then broke apart here, every day, for every production. Film after film. Decade after decade.

Sometimes the food was delivered to locations and to the stages as well—and out to the backlot. At 40 Acres, for some reason no one could ever remember, the aged Arab Village set, unless it was actually being filmed, was always the designated spot for dining on the backlot. Susan Myrick, a *Gone with the Wind* technical advisor, recounted how the old set "gives protection from wind that sometimes blows cold off the nearby mountains. A box that contains sandwiches, a hard-boiled egg, a carton of salad, an orange, and a slice of cake is given to everyone and there is steaming hot soup for all who wish it and milk and coffee too."[6]

One of the films that used the Arab Village as a commissary, and as a location, was the gargantuan *Greatest Story Ever Told* (United Artists 1965). When they were shooting in the stages, they used the Commissary itself, begrudgingly. A memo from George Stevens Productions' Eric G. Stacey to Desilu complained explicitly about "the crummy, so-called café on the Culver lot." Stacey went on to carp that "the present café is entirely inadequate in size, the service is almost non-existent, and the quality of the food provided is awful, as well as being unappetizing in appearance, and the manner in which it is served."

Today the fare is fancier, both on the sets and in the restaurant, which has been renamed, variously, the Plantation Club, the Eco Caters Café, and the Homegirl Café in recent years. Although they still cater meals out to the stages and to locations,

FRONT LOT: SETS, SETTINGS, AND SET PIECES   73

Sadly, little is left of the storied and sensational property department today, although fortunately these hand props from *King Kong*, seen here in 1989, survive, carefully preserved in the Sony Studios Archives.

12. Mail Room, First Aid, and Security (later Stages 5 and 6)

The current Stages 5 and 6 were built here in 1987, but the real estate these stages stand on has a varied history indeed.

Originally, and as noted earlier, these were Stages 2 and 3, each 72 by 180 foot, glass, "daylight" stages, built in 1919 and 1920, respectively. Both building were, in 1926, moved slightly east and aligned with the Van Buren Place wall. In 1929 they were renumbered as Stages 5 and 6, although they would only occasionally be utilized for production after that.

During the RKO and Selznick periods, Stage 5 was occasionally referred to as "Old Stage 5," although it housed props, wardrobe, and special-effects camera equipment. At the time of our survey, Stage 5 was still officially identified as the "Camera Effects Building"—although it was actually being used as the home of the security department, the mail room, first aid, and whatever else might have needed a few feet of studio real estate.

Stage 6, behind Stage 5 was, at this time, a nearly identical building often referred to as Building 6. It should not be confused with an earlier Stage 6, which

The original Stages 2 and 3, later Stages 5 and 6, have received so little respect over the years that they are only peripherally in this photo, far right. (1937)

FRONT LOT: SETS, SETTINGS, AND SET PIECES

This is a photo of the rafters above the original Stages 5 and 6, or perhaps it's the rafters above Stage 1. Take your pick. (1924)

burned down in 1927. *This* Building 6 was, in the late 1950s, stocked with grip, electrical, plumbing, and machine shop paraphernalia. In the 1960s the "alley" between these "stages" was utilized in *The Green Hornet* (TV 1966–67) for a stock-shot chase, which ultimately showed up in the series again and again.

The Grant Tinker / Gannett overhaul ended this era in 1987 by demolishing both buildings and constructing two entirely new stages, still called Stages 5 and 6, which were designed, as per Tinker's small-screen background, for television/audience shows. Building P, between the two stages, was designed with this in mind as

The current Stages 5 and 6, as they looked in 2015.

78 HOLLYWOOD'S LOST BACKLOT

well, with control booths and dressing rooms that could cannily be utilized by either stage. The entrances to these stages were aptly designed with colonial-style awnings and columns to reflect the much-older architecture on display at the Mansion and at some of the nearby bungalows.

As noted, today's Stages 5 and 6 did not exist at the time of our survey. But the current Stages 5 and 6 have been utilized, however, for the following:

Stage 5 (102 × 131 feet, 13,362 square feet)

Over the Top (Warner Bros. 1987): Sylvester Stallone arm wrestles here and on Stages 10 and 15.

Raising Miranda (TV 1988): short-lived sitcom with James Naughton.

My Wildest Dreams (TV 1995): short-lived sitcom with Lisa Ann Walter.

Love and Marriage (TV 1996): short-lived sitcom with Patricia Healy.

Malcolm & Eddie (TV 1996–2000): pilot only, apparently.

Life . . . and Stuff (TV 1997): short-lived sitcom with Pam Dawber.

Over the Top (TV 1997): short-lived sitcom with Tim Curry. Not to be confused with the Sylvester Stallone movie shot on the same stage.

Hush (Columbia 1998): Jessica Lange thriller may or may not have shot here. The production did rent the stage, for something.

Working Girl (TV 1990): a young Sandra Bullock in another short-lived sitcom.

WIOU (TV 1990–91): short-lived series set in a TV station.

Malice (Columbia 1991): assorted scenes shot here. Teamed Nicole Kidman and Alec Baldwin under the working title "Damages."

Between Brothers (TV 1997–99): pilot only.

The Gregory Hines Show (TV 1997–99): sitcom that brought the noted actor/dancer to the small screen.

Sex Court (TV 1998–2002): Julie Strain hosted this Playboy TV staple, which shot here circa 2001.

Live from Baghdad (TV 2002): Michael Keaton vehicle shot assorted scenes here.

All Grown Up (TV 2003): assorted scenes.

Arrested Development (TV 2003–): first season.

Misconceptions (TV 2006): six episodes shot here.

Mr. Woodcock (New Line 2007): assorted scenes; also shot on Stages 10 and 12.

The Ricki Lake Show (TV 2012–14): a live-audience daytime talk show.

Stage 6 (102 × 42 feet, 14,482 square feet)

Pee-wee's Playhouse (TV 1986–91): legendary children's show, which just maybe wasn't really for children.

A Family for Joe (TV 1990): Robert Mitchum, many miles and many years away from his film noir past, returns to the studio in a TV movie brimming with adorable orphaned children, which briefly spanned a sugary sitcom.

Step by Step (TV 1991–98): early episodes only, after which the soon-to-be-popular show moved to Warner Bros.

The Nanny (TV 1993–99): sitcom; Fran Drescher was the nanny.

Bette (TV 2000–1): sitcom; Bette Midler was Bette.

Life with Bonnie (TV 2002–4): sitcom; Bonnie Hunt was Bonnie.

Regular Joe (TV 2003): five-episode sitcom; Daniel Stern was Joe.

The Amazing Westerbergs (TV 2004): assorted scenes.

Deal or No Deal (TV 2005–9): game show, hosted by Howie Mandel.

STAGES 7, 8, AND 9

Stages 7, 8, and 9 were probably built by DeMille in 1928, although construction dates from different sources seem to waver between 1926 and 1930. There is a 20-foot-wide door between Stage 7 and Stage 8, and another between Stage 8 and Stage 9, so again, some projects listed below shot in two or in all three stages using connected sets.

13. Stage 7 (80 × 70 feet, 30 feet high, 5,600 square feet)

Bare Knees (RKO 1931): short subject that also shot on Stages 2 and 3 and 40 Acres.

Not So Loud (RKO 1931): short subject that also shot on Stage 9 and 40 Acres.

She Snoops to Conquer (RKO 1931): short subject that also shot on Stage 9 and 40 Acres.

Thanks Again (RKO 1931): short subject, shot entirely here.

The Big Gamble (RKO 1931): shot here and on Stages 1, 2, 3, 4, and 8 and 40 Acres.

Stages 7, 8, and 9, in 1934 (11, 12, and 14 are on the right), and in 2017.

The Past of Mary Holmes (RKO 1932): assorted scenes.

The Saddle Buster (RKO 1932): assorted scenes; also shot on 40 Acres.

Panama Flo (RKO 1932): Robert Armstrong, soon to costar with King Kong, here works instead with Helen Twelvetrees. Also shot on Stages 8, 10, and 14 and 40 Acres.

A Star Is Born (United Artists 1937): The first of many versions of the rising star / falling star chestnut. Assorted scenes shot here and around the lot, as noted below.

Gone with the Wind (MGM 1939): Bits and pieces of film were exposed here, including some involving Gerald O'Hara's death and a few Tara interiors.

Intermezzo: A Love Story (United Artists 1939): "taxi."

Citizen Kane (RKO 1941): "Int. Cellar," "great hall," "Thatcher Boardroom" (the same set as on Stage 4), "NY breakfast room," "Bernstein's office," "Madison Square Garden," "Kane's NY breakfast room," and "street."

Syncopation (RKO 1942): Most of this intriguing, generation-spanning look at popular music was shot here and on Stages 3, 9, 12, and 15.

The Fallen Sparrow (RKO 1943): shot here and on Stages 8, 11, and 14.

The Seventh Victim (RKO 1943): "Subway entrance" and "2nd Avenue." This unusual Val Lewton horror movie also shot on Stages 8 ("Jacqueline's room and upper hall"), 11 ("Highcliff Academy"), and 15 ("Bellevue hospital morgue").

A Night of Adventure (RKO 1944): "courtroom and corridor" was here. "Erica's room" was next door on Stage 8.

Honeymoon (RKO 1947): a grown-up Shirley Temple, here and on Stages 3, 4, 11, 12, and 15.

Portrait of Jennie (SRO 1948): "coffee shop."

Every Girl Should Be Married (RKO 1948): "Int. Cellar, Department Store."

Mighty Joe Young (RKO 1949): the "nightclub basement" where Joe is kept. Other stages this one shot in are detailed below.

A Woman's Secret (RKO 1949): Maureen O'Hara is the woman. Shot here and on Stage 8.

White Tower (RKO 1950): The "top of chimney" scenes were actually shot on the roof of this stage. Mountain exteriors were on Stage 15.

The Real McCoys (TV 1957–63): assorted scenes; also shot at various times on Stages 8, 9, and 14.

Carrie (United Artists 1976): One of the most famous thrillers of the '70s shot here and on Stage 8.

Cracking Up (Warner Bros. 1983): Jerry Lewis in a late-career comedy, shot as "Smorgasbord."

Against All Odds (Columbia 1984): inserts only.

Fright Night (Columbia 1985): assorted scenes; also shot on Stages 8, 9, and 15.

Cobra (Warner Bros. 1986): assorted scenes shot here and on Stage 8.

Peggy Sue Got Married (Columbia 1986): rehearsals only.

House Party (New Line 1990): assorted scenes shot here and on Stage 11.

Bugsy (TriStar 1991): assorted scenes; also shot on Stages 8, 9, 12, and 15.

Don't Tell Mom the Babysitter's Dead (Warner Bros. 1991): assorted scenes.

Ruby Cairo (Miramax 1992): assorted scenes shot here and on Stage 8.

Payback (Vidmark Entertainment 1995): direct-to-video C. Thomas Howell thriller, here and on Stage 8.

The Rock (1996 Disney): motion control effect shots.

Legally Blonde (MGM 2001): many days of reshoots, here and on Stage 8.

Repli-Kate (20th Century Fox 2002): straight-to-video feature shot here and on Stage 8.

14. Stage 8 (80 × 70 feet, 29 feet high, 5,600 square feet)

The Common Law (RKO 1931): assorted scenes; also used Stages 3, 4, and 7.

What a Time (RKO 1931): assorted scenes; also used Stage 9 and 40 Acres.

Lady with a Past (RKO 1932): assorted scenes; also shot on Stages 2, 4, 7, 9, 11, 12, and 14.

The Adventures of Tom Sawyer (United Artists 1938): the schoolmaster's home.

Gone with the Wind (MGM 1939): Inserts, close-ups, and pick-ups were shot on this stage; for example, the letter Scarlett gets informing her of the death of her first husband and Rhett's letter returning her and Melanie's donated-to-the-cause wedding rings. Also, the lumber mill interiors were shot on this stage, as well as many of the wardrobe stills used for reference by the costume department and some "road on edge of town" and the mysterious, unused "Provost Marshal" scenes (here and on Stage 9).

Intermezzo: A Love Story (United Artists 1939): The interiors set in a café and a camera shop were recorded here, as were some train interiors. Selznick sent a 1939 memo regarding these interiors complaining that "there is apparently something wrong with the interior of carriages, cars, etc. One side seems to be lower than the other with the result that the person we want to look short looks tall and vice versa." He indelicately warned about the tall (reportedly 5 feet, 9 inches) Ingrid Bergman "looking like a giantess."

Citizen Kane (RKO 1941): "Kane's boarding house," "Georgie's place," "Chicago Hotel Room," the "breakfast room," "Bernstein's office," and "Ext. Inquirer Bld" are among the odds and ends crafted here.

The Saint in Palm Springs (RKO 1941): as Palm Springs, presumably; also shot on Stages 9 and 12.

None but the Lonely Heart (RKO 1944): Cary Grant, playing against type, got an Oscar nomination for this one, which shot here and on Stage 15 for forty-one days.

Swedish import Ingrid Bergman looks petite in front of her bungalow, but in front of most of her American leading men, her height would prove to be more of a problem. (1939)

FRONT LOT: SETS, SETTINGS, AND SET PIECES

Marine Raiders (RKO 1944): The marines also raided Stages 9, 14, and 15.

Criminal Court (RKO 1946): The criminal court was here, although the corridors outside that court were shot on Stage 9.

Born to Kill (RKO 1947): assorted noir scenes.

Trail Street (RKO 1947): "Susan's home." The film also shot on Stages 14 ("Ferguson's Ranch, process") and 11.

Riff-Raff (RKO 1947): assorted scenes; also shot on Stages 11 and 12.

A Woman's Secret (RKO 1948): assorted scenes; also shot on Stage 7.

The Falcon's Adventure (RKO 1948): assorted dingy offices and police station interiors; other scenes shot on Stages 11-12 and 15.

They Live by Night (1948): The "cheap nightclub and washroom," "cabin hideout," and "Border Town" sequences shot onstage here by director Nicholas Ray.

Bodyguard (RKO 1948): "Hollywood ball park" and homicide department offices; also shot scenes on Stage 2.

Mighty Joe Young (RKO 1949): "Max's office." Max was played by Robert Armstrong, reprising his Carl Denham role from *King Kong*, sort of.

Adventure in Baltimore (RKO 1949): assorted scenes; also shot on Stages 7, 8, 12, and 15 and 40 Acres.

The Company She Keeps (RKO 1951): as "Gilmore Stadium," an actual ballpark in Hollywood, which would be demolished the following year to build CBS Studios.

The Blue Veil (RKO 1951): assorted scenes; also on Stages 2 and 3.

The Real McCoys (TV 1957–63): 1963 episode "Up to Their Ears in Corn" and others.

When Every Day Was the Fourth of July (TV 1978): assorted scenes; also, maybe, used Stage 9.

Those Lips, Those Eyes (United Artists 1980): assorted scenes.

Dial M for Murder (TV 1981): The remake of Hitchcock's classic was shot here, along with Stage 9.

True Confessions (United Artists 1981): assorted scenes; also shot on Stage 11.

City Heat (Warner Bros. 1984): Clint Eastwood and Burt Reynolds teamed here and on Stage 11.

Patty Hearst (Atlantic Releasing Corporation 1988): assorted scenes; also shot on Stage 11.

Sunset (TriStar 1988): This 1929-set Hollywood adventure was shot here and on Stages 9 and 11. James Garner played Wyatt Earp; Bruce Willis was Tom Mix.

Jake Spanner: Private Eye (TV 1989): Robert Mitchum puts on his well-worn trench coat one more time.

The Gambler Returns: The Luck of the Draw (TV 1991): Kenny Rogers saddles up for the fourth time in the title role. Also shot on Stage 8.

The American President (Universal 1995): assorted scenes shot here and on Stage 14-15.

The Cable Guy (Columbia 1996): rehearsals only, apparently.

Jerry McGuire (TriStar 1996): assorted scenes.

Up Close & Personal (Disney 1996): A 1997 book, *Monster: Living Off the Big Screen* by Gregory Dunne, describes the eight-year(!) odyssey to get this successful film made, which finally happened here and on Stage 16.

Action (TV 1999–2000): set in Hollywood; shot here and on Stage 7.

The Lot (TV 1999–2001)—again, set in Hollywood and also shot on Stage 7.

15. Stage 9 (80 × 70 feet, 30 feet high, 5,600 square feet)

Against the Rules (RKO 1931): short subject; also shot on Stage 3 and 40 Acres.

Oh, Marry Me (RKO 1931): short subject.

Sweepstakes (RKO 1931): assorted scenes.

Young Bride (RKO 1932): boat scenes.

Secrets of the French Police (RKO 1932): "Dorain's tenement" was here. The film also shot on Stage 3.

Nothing Sacred (United Artists 1937): "Ext. East River (packing box sequence)" shot here. "Hazel's hotel" was over on Stage 14.

Gone with the Wind (MGM 1939): The blink-and-you'll-miss-them shots of Gerald and Ellen O'Hara's grave markers were shot here. Also, more scenes involving the mysterious "Provost Marshall," ultimately cut, were filmed here.

Rebecca (United Artists 1940): "Dr. Barker's, night."

Bar Buckaroos (RKO 1940): assorted scenes shot here and on 40 Acres.

The Fargo Kid (RKO 1940): The "saloon" was here, although the sheriff's office was on Stage 3. Scenes for this movie were also shot at the RKO and Iverson ranches. The Kid was played by Tim Holt.

The Magnificent Ambersons (RKO 1942): Scenes involving the "veranda arbor," the railroad station, the courthouse, and the ballroom, as well as what is described as "Int. Runabout (process)" scenes, were composed here.

Behind the Rising Sun (RKO 1943): "Jap headquarters" is what the daily production reports crudely say was shot on this stage. The expansive film also shot on Stages 2, 3, 4, 12, and 15 and out on 40 Acres.

China Sky (RKO 1945): Dr. Thompson's (Randolph Scott) room; also process shots here and on Stage 8. Stage 14 was also used for prison scenes.

The Long Night (RKO 1947): Joe Adams's (Henry Fonda) room, "stairway and landing," etc. Exteriors looking through, or at, a window were actually shot outside the stage. Also see Stage 15.

The Company She Keeps (RKO 1951): assorted scenes; also shot on Stage 8.

Flying Leathernecks (RKO 1951): assorted scenes; also shot on Stages 3, 4, 7, 12, 14, and 15.

My Forbidden Past (RKO 1951): assorted scenes; also shot on Stages 2, 3, 7, 11, and 12.

Father Murphy (TV 1981–83): assorted scenes; also shot on Stage 15.

Ripley's Believe It or Not! (TV 1982–86): movie tough guy Jack Palance hosted; also shot on Stage 12.

Captain EO (Disney 1986): A studio medic had to be summoned when Michael Jackson injured his hand shooting a stunt on this stage for the long-running (1986–96) Disney theme park 3D ride, directed by Francis Ford Coppola and produced by George Lucas. Also shot on Stage 15.

Extreme Prejudice (TriStar 1987): assorted scenes; also shot on Stages 14, 15, and 16.

Monster Squad (TriStar 1987): cult comedy-thriller, at least among people who were of a certain age in 1987. Also shot on Stage 16.

The Kindred (FM Entertainment 1987): Rod Steiger starred; also shot on Stage 2.

RoboCop (Orion 1987): assorted scenes.

Beauty and the Beast (TV 1987–90): assorted scenes; also shot on Stage 14.

Beetlejuice (Warner Bros. 1988): assorted scenes. Tim Burton's blackly comic classic also shot on Stages 11, 12, and 14.

Ghosts of Mississippi (Columbia 1996): assorted scenes; also shot on Stage 15.

Red Corner (MGM 1997): Most this Chinese-set drama was shot here and in other LA locations. The runway model scenes were filmed on this stage; the courtroom was on Stage 15, and a house set was on Stage 14.

Stuart Little (Columbia 1999): assorted scenes; also shot on Stages 14, 15, and 16.

The L Word (TV 2004–9): 2004 pilot only.

Yours, Mine & Ours (Columbia 2005): assorted scenes; also shot on Stages 14 and 15.

Monster-in-Law (New Line 2005): assorted scenes; also shot on Stages 10, 12, and 14.

Stranger Than Fiction (Columbia 2006): "Ana's bedroom."

Beowulf (Paramount 2007): rehearsals mostly, here and on Stage 10.

STAGES 11, 12, AND 14

In 1927 an electrical short-circuit is believed to be responsible for a fire that completely destroyed the old Stage 6 and an elaborate canvas-covered ship set constructed for *The Wreck of the Hesperus* (Pathé 1927). Undaunted, producer Cecil B. DeMille immediately built a new, much larger structure on top of the ashes.

Stage 11, later Stages 11, 12, and 14, under construction. (1927)

The new building was, at the time, the largest film stage in the world. A. G. Volk, DeMille's assistant manager, excitedly, but not entirely accurately, told the press that the structure was "large enough to accommodate a football field, with room to spare . . . An airplane could take off from within its confines."[7]

Although it has long been divided into Stages 11, 12, and14 (superstition prohibits a Stage 13 at any Hollywood studio), and although the structure was largely rebuilt in 1991, and it has since been supplemented by Stages 15 and 16, the building still stands today.

In the late 1920s, the words "Pathé Studios Culver City" were painted onto the roof, reportedly by a local Girl Scout troop! Those girls must have done a good job as the sign survived on that very long, high rooftop for decades.

16. Stage 11 (98 × 136 feet, 40 feet high, 13,328 square feet)

The Main Event (Pathé 1927): The inaugural film shot on the big stage, one of the few made before it was divided into three sections.

Sunny (RKO 1930): The theater scenes, front and backstage, were shot here and on Stage 12, although the stage door scenes outside that theater were ultimately shot on Stage 7.

The Penguin Pool Murder (RKO 1932): assorted scenes.

Ann Vickers (RKO 1933): assorted prison sets.

Aggie Appleby, Maker of Men (RKO 1933): assorted scenes; also shot on 40 Acres.

Snug in the Jug (RKO 1933): a (Bobby) Clark and (Paul) McCullough comedy short; "jail" and "interior limousine" scenes.

A Star Is Born (United Artists 1937): This stage was a catch-all for many scenes in this early version of the oft-filmed Hollywood fable. The railroad station where Esther (Janet Gaynor) says good-bye, the Sunset Strip's Trocadero nightclub, and the Malibu dream house, among other sets, were all here.

The Prisoner of Zenda (United Artists 1937): the coronation scene and railroad station set.

The Adventures of Tom Sawyer (United Artists 1938): "Int. Sunday School," the church, and Tom's home were all here.

Gone with the Wind (MGM 1939): All of Twelve Oaks plantation was built here, with the set's doorway constructed in the actual doorway of the stage's north wall so that the camera could glide from an exterior to an interior in a single shot, which it does. Also, Rhett and Scarlett's Atlanta mansion and second-floor interiors, including daughter Bonnie Blue Butler's room, were here.

The great hall of Twelve Oaks for *Gone with the Wind* was built so that its doorway corresponded to the actual doorway of Stage 11, so the entrance seen here in back of Vivien Leigh and a lot of costumed extras actually looks outside toward Stage 7, itself costumed with a painted backdrop. (1939)

Intermezzo: A Love Story (United Artists 1939): "Int. Stockholm station."

Rebecca (United Artists 1940): Screen tests for assorted actresses vying for the lead in this much-anticipated production were shot here in the "Rhett Butler Bedroom" set still standing from *Gone with the Wind*. Anne Baxter, Loretta Young, and Margaret Sullavan were among the hopefuls. Vivien Leigh, who wanted very much to work with lover Laurence Olivier, read for the part with Olivier, but Selznick still passed. Later in the production, eventual leads Olivier and Joan Fontaine shot beach and boathouse scenes, as well as "Maxim's suite" and "Hotel De Paris" sequences, on the same stage in a new set.

Tom Brown's School Days (RKO 1940): assorted scenes; also shot on Stage 12 and 40 Acres.

Stranger on the Third Floor (RKO 1940): Peter Lorre menaces here and on Stage 4.

FRONT LOT: SETS, SETTINGS, AND SET PIECES

Citizen Kane (RKO 1941): Susan's apartments and rooms, the "drugstore and street exteriors," part of the Madison Square Garden sequence, "ext. observation car," and Kane's assorted offices were built here. Also, tests were shot here for the "boat dock" and the "everglades tent," the latter of which includes some strange animated birds left over from *King Kong*. Critic Roger Ebert once wrote about the tent scene that "if you look closely there seems to be a pterodactyl flapping its wings."[8] Look for it.

The Devil and Miss Jones (RKO 1941): Some of the Xanadu mansion sets from the recent *Citizen Kane* show up here, and on Stage 12 as well.

The Devil and Daniel Webster (RKO 1941): Process shots of "Belle's carriage" were filmed here and on Stage 12. Other scenes for one of RKO's most audacious releases were shot on Stages 3, 4, 7, 9, 14, and 15 and outside at 40 Acres.

The Magnificent Ambersons (RKO 1942): The Amberson home sets were largely constructed here.

Cat People (RKO 1942): "Irena's apartment" and "Oliver's apartment." Some of the sets were *Magnificent Ambersons* stock units, repurposed.

Rookies in Burma (RKO 1943): the "Burma tavern," of course.

Passport to Destiny (RKO 1944): train and ship interiors.

I'll Be Seeing You (United Artists 1944): The prison, railroad coach, dress shop, and YMCA entertainment room were here. Stages 2, 4, and 7 (the latter identified as used for "dogfight" scenes) were also utilized.

Since You Went Away (United Artists 1944): The poignant, and much parodied, railroad station good-bye scene between Jennifer Jones and Robert Walker was shot here, as were the bowling alley scenes, the hotel cocktail bar, the country road, and part of the aircraft hangar dance number.

The Fighting Generation (RKO 1944): This short was shot on a hastily constructed hospital ward set on Saturday, October 7, 1944, by director Alfred Hitchcock to raise money for government war debts. Hitchcock, stars Jennifer Jones and Rhonda Fleming, producer Selznick, and cinematographer Gregg Toland all patriotically donated their time.

Spellbound (United Artists 1945): "Peterson's office" and some more of the Salvador Dali dream scenes (involving a rooftop and chimney, and the mysterious "masked man") were shot here, as was the memorable, deliriously Freudian shot of an embrace between Gregory Peck and Ingrid Bergman, superimposed over dozens of ghostly doors unlocking and opening up, one behind the other, into infinity, all to Miklós Rózsa's swirling score.

It's a Wonderful Life (RKO 1946): "George's Dodge" is cryptically referred to as shooting here. As noted in the following pages, Frank Capra's masterpiece shot on other stages around the lot as well.

Woman on the Beach (RKO 1947): Jean Renoir directed this one, here and on Stages 2, 12, 14, and 15.

Mourning Becomes Electra (RKO 1947): "Wharf and shop-matte-fog" is how this set is described in the production paperwork. Other scenes for this one were shot on Stages 9 ("sitting room"), 14 ("Brandt's room"), and 15 ("manor house").

Mighty Joe Young (RKO 1949): "nightclub."

The Racket (RKO 1951): assorted scenes; also shot on Stages 7, 8, and 14.

His Kind of Woman (RKO 1951): shot here and on Stages 12, 14, and 15, with many Howard Hughes–ordered reshoots along the way.

One Minute to Zero (RKO 1952): assorted scenes; also shot on Stages 3, 7, 12, and 15.

Guestward Ho! (TV 1960–61): New Mexico–set sitcom.

Uptight (Paramount 1968): For this remake of John Ford's 1935 RKO classic, *The Informer*, director Jules Dassin updated the story from the 1922 Irish rebellion milieu of the original to the then-topical 1968 Cleveland, Ohio, race riots. The film shot on location for two weeks, then moved here, seamlessly blending forced perspective sets, miniatures, and translucent backings with actual locations. The warehouse and back alley set on this stage continued through the elephant doors and into Stage 12 as well. The film also shot on Stage 16, where an elaborate rooftop city view—involving miniatures and photo cut-outs of buildings—was constructed.

The Fortune (Columbia 1975): Mike Nichols's fascinating flop shot interiors here and exteriors nearby at 40 Acres.

The Jerk (Universal 1979): assorted scenes.

Gideon's Trumpet (TV 1980): Fay Wray (in her last role) returns to the lot she once ran away from King Kong in. Her leading man this time is the less-hirsute Henry Fonda.

Under the Rainbow (Orion 1981): assorted scenes.

Pennies from Heaven (MGM 1981): Steve Martin (*The Jerk*) returns to the stage, to eventual critical, if not popular, success.

Turbulence (MGM 1981): assorted scenes; also shot on Stage 3.

Best Friends (Warner Bros. 1982): Burt Reynolds and Goldie Hawn have the title roles.

Right of Way (TV 1983): Bette Davis and James Stewart, both age 74 in 1983, made their only on-screen appearance together here.

The Creature Wasn't Nice (Almi Pictures 1983): assorted scenes; also shot on Stage 12.

Bachelor Party (20th Century Fox 1984): assorted scenes.

Sledge Hammer! (TV 1986–88): well-remembered cop show spoof, shot here.

Masters of the Universe (Cannon 1987): Dolph Lundgren is well cast as He-Man. Shot here and over on Stage 12.

The Van Dyke Show (TV 1988): Dick Van Dyke, with his son Barry in tow, returned to television for this short-lived sitcom.

Spellbinder (MGM 1988): no relation to *Spellbound*; shot under the title "The Witching Hour."

Skin Deep (20th Century Fox 1989): assorted scenes.

Pacific Heights (20th Century Fox 1990): assorted San Francisco–area locations, here and on Stage 12.

Hook (TriStar 1991): Steven Spielberg's take on the Peter Pan myth shot the exteriors of Wendy's house here, as well as some footage on Stages 8 and 12 and on the old MGM (now Sony) lot.

City Slickers (Columbia 1991): assorted scenes.

A Few Good Men (Columbia 1992): assorted scenes; also shot on Stages 4 and 12.

Honeymoon in Vegas (Columbia 1992): assorted scenes.

Mad About You (TV 1992–99): The beloved sitcom also shot (only occasionally) on Stage 12.

Undercover Blues (MGM 1993): assorted scenes; also shot on Stage 16.

Can't Hurry Love (TV 1995–96): occasional.

Aladdin (Cannon 1996): assorted scenes for Miami-set(?) *Arabian Nights* update.

Contact (Warner Bros. 1997): Assorted scenes shot were here and on Stages 15 and 16. Stage 8 was used for plane interiors, and Sony Pictures Imageworks used Stages 2 and 3 for visual effects. The Warner Hollywood lot was also utilized.

Mr. Deeds (Columbia 2002): reshoots only.

Watching Ellie (TV 2002–3): Julia Louis-Dreyfus in a sitcom intriguingly told in real time, meaning that the audience and the heroine would discover things at the same rate, only with commercials.

Peter Pan (Columbia 2003): Peter Pan returns to the stage.

Christmas with the Kranks (Columbia 2004): assorted scenes.

For Your Consideration (Warner Bros. 2006): Christopher Guest mockumentary staged some studio scenes here.

17. Stage 12 (99 × 136 feet, 40 feet high, 13,464 square feet)

Prestige (RKO 1932): assorted scenes; also shot on Stages 3 and 4.

What Price Hollywood? (RKO 1932): Stage 12 was certainly well-cast as "movie studio set."

The Most Dangerous Game (RKO 1932): sets for both chateau and jungle scenes.

The Great Jasper (RKO 1933): assorted turn-of-the-century settings.

King Kong (1933): Those *Most Dangerous Game* jungle sets, noted above, including a mossy log spanning a jungle ravine that was referred to internally as "Cooper's log," were memorably recycled here. *Kong* also shot on Stage 14, as detailed below.

The Right to Romance (RKO 1933): assorted scenes; also shot on 40 Acres.

Before Dawn (RKO 1933): assorted scenes.

Christopher Strong (RKO 1933): "Carrie's Villa—Cannes" and "Christopher's Pier" in Katharine Hepburn's second film.

Little Women (RKO 1933): Katharine Hepburn's fourth film shot on this stage for one day, on August 2, 1933.

The Great Jasper (RKO 1933): "Int. Silver hut (1916)."

The Right to Romance (RKO 1933): The Preble family Long Island home was here.

The Son of Kong (RKO 1933): Again, jungle scenes, and the "temple ledge and pool" were here. The "exterior temple, earthquake," however, was on Stage 11.

Hooray for Love (RKO 1935): "barracks and hall."

A Star Is Born (United Artists 1937): The "billboard montage" sequence, the "designer's office," the "Sunrise Court," the shabby "Oleander Arms" where Esther stayed, and "Niles' office" were here. Here as well was shot the famous "Ext. Chinese Theater" scenes, where hopeful starlet Esther admires the cement footprint impressions of Hollywood's immortals, as hopeful starlets still do today—actually, they were shot both here and at the real Chinese Theater. Incidentally, the construction cost to rebuild Hollywood's idea of China, onstage in Culver City, was $1,668.87—exactly $668.87 above the original estimate.

Gone with the Wind (MGM 1939): The dramatic Atlanta church/hospital scenes were shot here, including those involving the long-lived stained glass windows, which are partially shattered on-screen by a Yankee cannonball. Also, the Atlanta Bazaar, where Rhett and Scarlett first dance, was created here (by the first director, George Cukor), although the ceiling was actually supplemented by a beautiful matte painting.

Intermezzo: A Love Story (United Artists 1939): the Brandt home scenes and "Ann Marie's bedroom."

Made for Each Other (United Artists 1939): Miniatures, including those playing the Salt Lake City Airport, were shot here, as were some New York nightclub interiors. Other locations are detailed below.

Rebecca (United Artists 1940): "Manderley's gate and grounds" were here, although similar sets, full size and miniature, were on other stages. The late Rebecca's room as well as the room of the (unnamed in the book or movie) main character were also here.

Dance, Girl, Dance (RKO 1940): assorted scenes; also shot on Stage 3.

You'll Find Out (RKO 1940): "Ext. garden"; shot for one day on this lot, on this stage.

I'm Still Alive (RKO 1940): "Blake's office and outer office."

Mexican Spitfire Out West (RKO 1940): "Dennis's inner and outer office" and "Int. Skinner's."

Parachute Battalion (RKO 1941): The "Dummy village" was built here, but most sets were on Stage 15.

Forever and a Day (RKO 1943): multigenerational saga (with multiple directors and writers) shot here and on Stage 15 and out on 40 Acres.

Since You Went Away (United Artists 1944): "hotel lobby."

What a Blonde (RKO 1945): Interiors of "Fowler's office" were shot here, probably on a pre-standing available set; the rest of the movie was shot on the Gower lot.

The Story of G.I. Joe (United Artists 1945): "set No. 2, bivouac."

It's a Wonderful Life (RKO 1946): The brief "Ernie's taxi" scenes seem to be all this much-beloved picture shot on this particular stage.

Genius at Work (RKO 1946): process and rooftop shots.

Mighty Joe Young (RKO 1949): Much of the live-action footage for this follow-up to *King Kong* was shot on this stage, including "Jill's bedroom," the "Chinese restaurant," and much of the burning orphanage climax.

Little Women (MGM 1949): Nothing in the finished film was actually shot on this stage, but it is now known that in 1946 Selznick shot scenes here for a version of the classic novel which was never completed. For unknown reasons, the film would be shut down after several days of filming and the roles held by Jennifer Jones (who was to play Jo), Diana Lynn (Amy), and Rhonda Fleming (Meg) would be recast. Gregory Peck, Jones's costar in *Duel in the Sun* (SRO 1946), was also to have had a role. Selznick did eventually sell the production to MGM, which released its (recast) version to the world in 1949.

The Whip Hand (RKO 1951): shot here and on Stages 11 and 15 as "The Man He Found."

The Big Sky (RKO 1952): Many of the supposed exteriors were shot here and on Stages 12, 14, and 15.

Androcles and the Lion (RKO 1952): Assorted pastoral exteriors including the "bubbling brook" were re-created here and on Stage 11. The "tailor shop," however, was on Stage 7.

The Carol Burnett Show (TV 1967–78): Fans of this beloved variety show probably know that it shot at CBS Studios in Hollywood, but for a few episodes in the 1970s, during a strike at CBS, the show relocated here.

The Sophisticated Gents (TV 1981): assorted scenes.

Human Highway (Shaky Pictures 1982): Neil Young and Devo! This oddity also shot scenes at the Raleigh Studios in Hollywood.

Kenny & Dolly: A Christmas to Remember (TV 1984): taped here.

A Fine Mess (Columbia 1986): assorted scenes; also shot on Stages 7, 8, 9, and 12.

Three Amigos! (Orion 1986): The famous campfire sing-along scene was shot here. Other scenes were shot on Stages 4, 12, and 16.

Frank's Place (TV 1987–88): acclaimed, short-lived sitcom.

Mr. Saturday Night (Columbia 1992): assorted scenes.

First Time Out (TV 1995): short-lived (twelve episodes) sitcom.

Matilda (TriStar 1996): Between failed sitcoms, this feature shot here for two days only.

Guys Like Us (TV 1998–99): short-lived (thirteen episodes) sitcom.

Shasta McNasty (TV 1999–2000): short-lived (twenty-two episodes) sitcom.

The Tick (TV 2001–2): The popular comic book would later span a successful TV series, but this wasn't it. Also shot on Stage 14.

The Mind of the Married Man (TV 2002–3): assorted scenes; exteriors shot in Chicago.

Cheaper by the Dozen (20th Century Fox 2003): assorted scenes.

Bride for Sale (RKO 1949): The entrance to arena, police station, and museum office sets were here. Meanwhile, the "fish market" was on Stage 15, and 40 Acres is in there too.

The Lusty Men (RKO 1952): well-regarded rodeo-set drama shot here and on Stage 15.

The Conqueror (RKO 1956): "Int. Khan's palace and hall." The film also shot on Stage 15 and out on 40 Acres.

The Greatest Story Ever Told (United Artists 1965): Among the first sets built in Culver City for this biblical spectacular was "Machaerus Fortress," here and on Stage 15. Later in this very long, very eventful shoot, many other locations on the lot were also utilized, as noted below.

Star Trek (TV 1966–69): Used only for the two pilots, "The Cage" (1964) and "Where No Man Has Gone Before" (1965). Gene Roddenberry, *Trek*'s creator, did not like the studio, and these stages in particular. He bellyached to Desilu about soundproofing problems, pigeons, plumbing, and even bees(!) disrupting production. Robert Butler, director of "The Cage," got the same non-answers when he complained to the crew about the plagues of vermin swarming the stage and was informed nonchalantly that "this is an old studio. It's where they burned the city of Atlanta for *Gone with the Wind*."

"They should have burned this stage as well," Butler wryly retorted.[9]

Star Trek also used Stages 15 and 16. Stage 15 at the time had a trench along the floor, of unknown origin, which Roddenberry added to his long list of complaints about the studio, which perhaps led to *Trek*'s subsequently being taken at Paramount.

Don Kirshner's Rock Concert (TV 1973–80): Premore Videotape Productions leased this stage in the 1970s and shot several projects on the lot, including this one, intermittently, on this stage.

The Cliffwood Avenue Kids (TV 1977): another one from the makers of *Don Kirshner's Rock Concert*.

Cher . . . Special (TV 1978): Appearing with Cher was Dolly Parton.

The Buddy Holly Story (Columbia 1978): The stage here was dressed to look like the set for the Ed Sullivan TV show. The film itself also shot on Stage 3.

The Entity (20th Century Fox 1983): assorted scenes; also shot on Stage 15.

Starflight: The Plane That Couldn't Land (TV 1983): passenger cabin.

Gleaming the Cube (20th Century Fox 1989): assorted scenes.

Lock Up (TriStar 1989): This Sylvester Stallone thriller may have also shot scenes next door on Stage 15.

Life Stinks (MGM 1991): assorted scenes.

Plymouth (TV 1991): set on the moon, but shot here and on Stage 15 instead.

The Craft (Columbia 1996): assorted scenes.

What Planet Are You From? (Columbia 2000): also shot on Stage 15.

Riding in Cars with Boys (Columbia 2001): reshoots; also shot on Stage 15.

Stuart Little 2 (Columbia 2002): also shot on Stage 15.

The Ladykillers (Disney 2004): Tom Hanks's apartment as well as some exterior sets were here and on stage 15.

Fat Albert (20th Century Fox 2004): Bill Cosby's animated series becomes a live-action feature, which also shot on Stage 16.

STAGES 15 AND 16

The newest stages on the lot, at least until the 1980s, Stages 15 and 16 were built by Selznick/RKO in 1940, although they would not be divided (creating Stage 16) until the early 1960s. There is a 68-foot-wide doorway between Stages 14 and 15, with a 3-foot elevation difference between the two stages and nearly that much between

Awe-inspiring aerial look at Stage 15 (later Stages 15 and 16) under construction, far left. (1940)

Stages 11, 12, 14, 15, and 16 are housed in one gargantuan building and together constitute the largest designated filming space in the world. (2015)

Stages 15 and 16. The door between those two stages, adding to the chaos, is only 38 feet wide. When combined with Stages 11, 12, and 14, this creates the possibility for an awe-inspiring and unparalleled 75,080-square-feet shooting space—the largest purpose-built soundstage in the world—although the internal walls make it technically ineligible for the record.

The stages were originally designed and intended, by Selznick and by RKO, for large-scale and for special-effects sequences. The rafters above the stages reflect this aspect of their construction and, in fact, even in the computer-generated twenty-first century, these rafters still contain largely unused, heavy-industry reinforcements, gimbals, and rail track intended for practical effects.

19. Stage 15 (132 × 129 square feet, 43 feet high, 17,028 square feet)

My Life with Caroline (RKO 1941): restaurant scenes.

The Lady Takes a Chance (RKO 1943): a rare John Wayne romantic comedy shot here.

Tender Comrade (RKO 1943): shot on this stage for part of one day only.

Mr. Lucky (RKO 1943): one day only, with Cary Grant.

Hitler's Children (RKO 1943): the labor camp, as well as the German hill and airport.

The Falcon in Danger (RKO 1943): assorted scenes.

Flight for Freedom (RKO 1943): "Newark Airfield" and "Mines Field air race" scenes shot on the stage for seven days. "Mines Field" was the name for what later became Los Angeles International Airport.

The Sky's the Limit (RKO 1943): Fred Astaire dances on a trestle and on a Flying Tiger plane.

Isle of the Dead (RKO 1945): "grave," "house and tunnel," and "beach" scenes for another Val Lewton horror classic.

The Bells of St. Mary's (RKO 1945): Most of the convent school classrooms and offices and grounds were constructed here, although Stage 14 stood in for Bogardus's (Henry Travers) office and boardroom.

Back to Bataan (RKO 1945): Many Bataan exteriors were shot here. The film also utilized Stages 2 (Marine HQ), 7-8 (aquatic scenes), 11 ("Penthouse"), and 14 ("guerilla bivouac").

Spellbound (United Artists 1945): The none-too-convincing (to modern eyes) skiing scenes with Gregory Peck and Ingrid Bergman were shot here by director Alfred Hitchcock.

Cornered (RKO 1945): Most of the French village scenes were shot here. The film also shot on Stages 7 and 8.

The Enchanted Cottage (RKO 1945): Most of the titular cottage scenes were filmed here, although a few other scenes—"Oliver's cottage bedroom," "Thatcher's office," and "Price's bedroom, summer"—were shot on Stage 7.

Sister Kenny (RKO 1946): The Kenny family home scenes were constructed here. The film also shot on Stages 3, 7, 8, 9, and 14.

It's a Wonderful Life (RKO 1946): "Potter's home" and the "river" are tantalizingly referred to in the production paperwork as having shot here.

The Falcon's Adventure (RKO 1946): assorted scenes.

Bedlam (RKO 1946): another Val Lewton thriller, again shot mostly on this stage.

The Long Night (RKO 1947): Most of this film's shadowy film noir effects were created on this stage, including several of the supposed exteriors, such as the "cemetery and river bank," the foundry and mill, and the bus stop and highway. The "Skyroom Restaurant," however, was on Stage 12, and process shots involving a bus were shot on Stage 4. Also see Stage 9.

Out of the Past (RKO 1947): Not treated as anything particularly special in 1947, this velvet-black film noir has gathered a deserved cult following. "White's house," "meadow and forest—night," "Int. and Ext.—Cathy's cabin," "clearing," "Cathy's Mexican cottage," and "Meta's apartment" were all built here, cheaply and quickly, and stacked next to one another across the stage. The production later moved to Stage 14, where "Eels' apartment" and "Tillotson's apartment" were built, although there is apparently no character in the film with the name "Eels."

The Iron Major (RKO 1947): "Campos (bonfire)."

I Remember Mama (RKO 1948): Process shots involving the actors standing in front of prerecorded footage of a ferry building and a hilltop cemetery were shot here.

The Miracle of the Bells (RKO 1948): Again, hilltop cemetery scenes, just like in *I Remember Mama*, were shot on this stage. The film also shot on Stages 2, 3, 7, and 8.

Blood on the Moon (RKO 1948): camp scenes.

Berlin Express (RKO 1948): The title-referenced train interiors were here, but this one also shot on Stages 9, 11-12 (combined for a "department store" set), and 14.

Three Godfathers (MGM 1948): John Wayne, John Ford, and Merian C. Cooper's dress rehearsal for *The Searchers* was on this stage and Stage 4.

Mighty Joe Young (RKO 1949): The furniture van, which Joe escapes from town in. Has it been mentioned that Joe is an oversize gorilla?

Holiday Affair (RKO 1949): Robert Mitchum romances Janet Leigh on a snowy Central Park set.

Roughshod (RKO 1949): "Ext. campsite" scenes were shot here for several days. Later in the production the "Forest trail" process shots were recorded here as well.

The White Tower (RKO 1950): assorted mountain exteriors.

Born to Be Bad (RKO 1950): assorted scenes. Improbably, good girl Joan Fontaine has the title role.

At Sword's Point (RKO 1952): "Claudine's Cottage," the "Golden Cockerel," and some odds and ends. Most of the film was shot on Stages 8 and 11-12 in 1949–50 as "Sons of the Musketeers."

Macao (RKO 1952, filmed in 1950): "Ext. Macao Waterfront."

Blackbeard, the Pirate (RKO 1952): Many of this grisly film's exteriors were shot onstage here, including the severed head on the plaza gates and Blackbeard's (Robert Newton, of course) historically ludicrous, but unforgettable, death scene, which Stephen King later plagiarized in *Creepshow* (Warner Bros. 1982). The cabin scenes were shot over on Stage 12.

The Searchers (Warner Bros. 1956): Another of the great American movies arrived at the studio in July 1955, after director John Ford's customary Monument Valley location work. Apparently, all of the soundstage scenes were created here, and on Stage 14, over the course of that month and into August, specifically the interiors at the ranch house where John Wayne's heroic and terrifying Ethan Edwards is introduced, and where his niece is kidnapped by Comanche braves. Other interiors, mostly detailing assorted farms and forts, were re-created on the stage as well.

Less successfully, several exteriors were created in Stage 15, including some campfire sequences which *New York Times* critic Bosley Crowther nitpicked in his review of the movie "could have been shot in a sporting-goods store window. That isn't like Mr. Ford. And it isn't like most of this picture, which is as scratchy as genuine cockleburs," Crowther carped.[10]

The Searchers was a box office success in 1956, but its reputation and its influence—like that of *King Kong, Citizen Kane, Fantasia, It's a Wonderful Life, Out of the Past, The Night of the Hunter, Star Trek, Raging Bull, E.T. the Extra-Terrestrial, The Matrix,* and other films crafted in this studio—has grown, and grown more legendary, in subsequent years.

The Girl Most Likely (RKO 1958): one of the last RKO pictures. Director Mitchell Leisen liked to say, perhaps apocryphally, that the sets were destroyed behind him by a wrecking crew as he finished with them.

The Untouchables (TV 1959–63): the 1962 episode "An Eye for an Eye," among others.

Batman (TV 1966–68): stately Wayne Manor.

Rich and Famous (MGM 1981): George Cukor, (briefly) the director of *Gone with the Wind*, returned here to the lot for what would turn out to be the 81-year-old's last project.

Escape from New York (1981 Embassy Pictures): process shots only.

A Streetcar Named Desire (TV 1984): Ann-Margret does a good job relying on the kindness of strangers, here and on Stage 16.

Maxie (Orion 1985): rehearsals only.

The Morning After (20th Century Fox 1986): assorted scenes. Star Jane Fonda would, years later, shoot *Monster-in-Law* (New Line 2005) on the same lot.

Tango & Cash (Warner Bros. 1989): Sylvester Stallone and Kurt Russell flex their considerable muscles here and on Stages 7, 9, 14, and 16.

Chances Are (TriStar 1989): assorted scenes.

Out on a Limb (1992 Universal): assorted scenes; also shot on Stage 9.

City Slickers II: The Legend of Curly's Gold (Columbia 1994): assorted scenes.

Nixon (Disney 1995): ballroom scenes with Anthony Hopkins in the title role.

The Story of Us (Columbia 1999): "Us" was Bruce Willis and Michelle Pfeiffer.

8mm (Columbia 1999): "Int. House" sets; also shot on Stage 14.

What Women Want (Paramount 2000): assorted scenes; also shot on Stage 16.

Joe Dirt (Columbia 2001): reshoots only.

S.W.A.T. (Columbia 2003): helicopter-set reshoots only.

Kill Bill: Vol 1 (Miramax 2003): assorted scenes, including Uma Thurman's memorable crawl out of the grave.

50 First Dates (Columbia 2004): assorted scenes.

Escape Plan (Summit 2013): Prison and cell sets were constructed here for the historic, muscle-bound, and momentous first-ever on-screen team-up of Sylvester Stallone and Arnold Schwarzenegger, who had first worked on the lot in 1976 and 1980, respectively.

20. Stage 16 (131 × 129 square feet, 46 feet high, 16,899 square feet)

The Greatest Story Ever Told (United Artists 1965): Art director William J. Creber was a young man with mostly television credits on his résumé when Richard Day, his boss, resigned from the film midway through the shoot. "I thought George Stevens was going to fire me too," Creber remembered. "Instead he gave me the picture." That picture would be in production for a year and a day, first in Utah and Arizona and later at 40 Acres and on this stage, among many others. "One day I was showing him [Stevens] the set," Creber recalls. "It was dug into a hillside on the soundstage, it had a cave, a sort of crevice, and Stevens kept looking at it. Finally, he sort of nodded to himself, and said 'I could see Stan and Ollie having a good time on this set.'"[11]

The great filmmaker had by this time been nominated five times for the Academy Award for Best Director, winning twice. He had also been the recipient of the

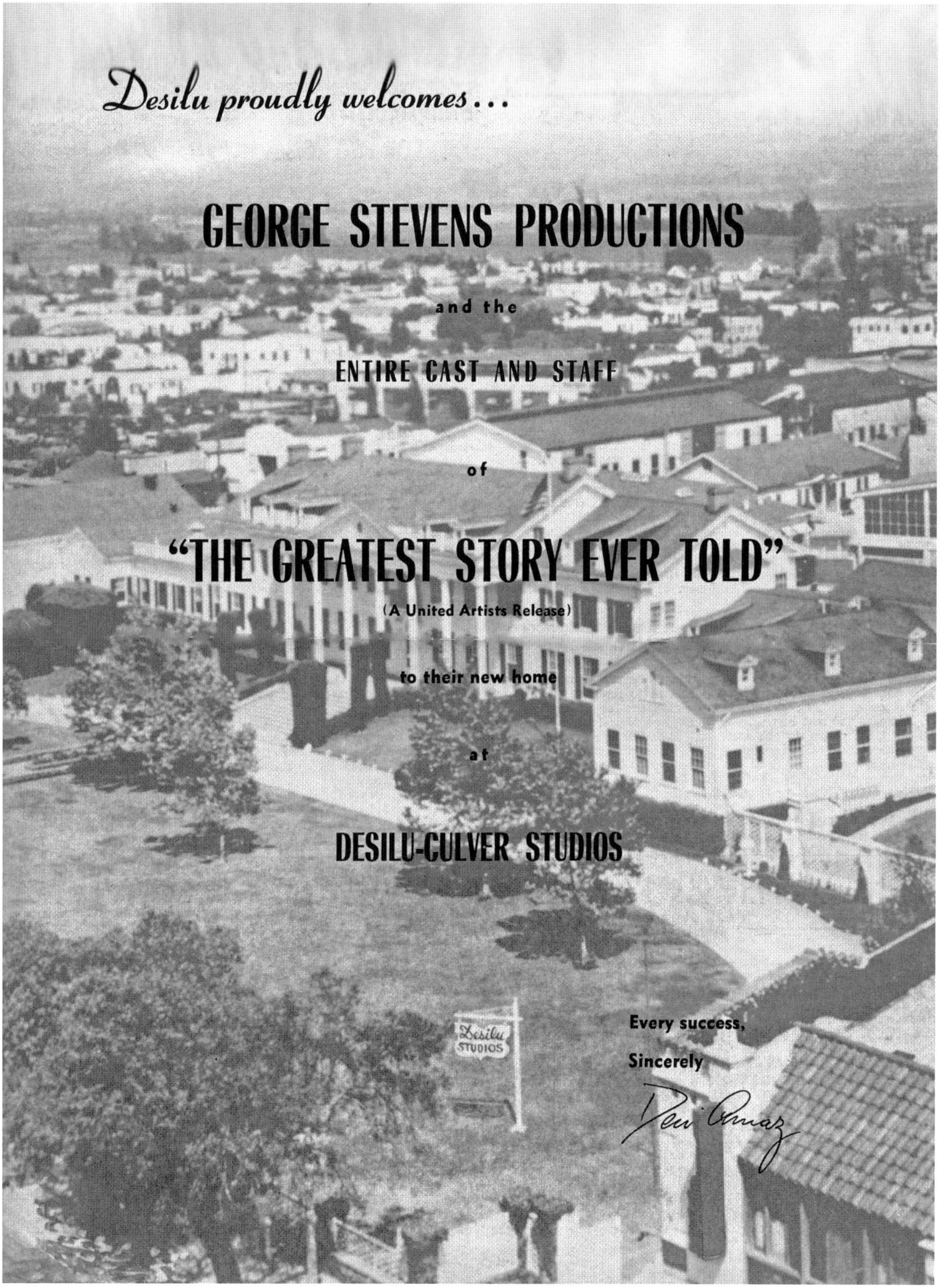

Desilu proudly welcomes...

GEORGE STEVENS PRODUCTIONS

and the

ENTIRE CAST AND STAFF

of

"THE GREATEST STORY EVER TOLD"

(A United Artists Release)

to their new home

at

DESILU-CULVER STUDIOS

Every success,
Sincerely

Desi Arnaz

The Greatest Story Ever Told came to Culver City with much fanfare, as shown in this trade ad, but the relationship between the studio and director George Stevens, like everything about the costly production, would prove to be difficult. (1965)

Irving G. Thalberg Memorial Award, the Directors Guild of America Best Director Award (three times), and the D. W. Griffith Award. He was also a former Academy of Motion Picture Arts and Sciences president. Yet on this day, at this unguarded moment, on this empty stage, rather like Charles Foster Kane dreaming of Rosebud, George Stevens was apparently looking back to his long-ago days as a camera operator and gagman for Laurel and Hardy.

Later in the production, the long march of Jesus (Max von Sydow), agonizingly dragging his cross to Mount Calvary is depicted here and on Stage 15. This is the sequence that incurred the wrath of critics, because of the admittedly distracting cameos by stars who aid, or jeer, Jesus along his route. (For example, Sidney Poitier briefly steps out of the crowd to help Christ with his burden.) The long, torturous sequence that follows is actually one of the most realistic portrayals of the crucifixion ever depicted. For the first time on screen, Mount Calvary is depicted as being a small knob of a hill just outside the city's back gate on the edge of the town dump. Leave it to an artist like Stevens, misguided though he often may have been here, to stage the climax to the "greatest story ever told" on a trash heap.

Batman (TV 1966–68): The Batcave, the wall for the climbing scenes, and Commissioner Gordon's office were all here. When the series was canceled by ABC in 1968, another network offered to pick up the show, but reneged when it was discovered that those expensive sets had already been removed from the stage.

Airplane (Paramount 1980): The aviation scenes were shot here using TWA's Airline Film & TV Promotions Boeing 707 cockpit and fuselage. Fans of the film will remember the flapping vultures that had to be rented and trained to sit in the set for one memorable gag. The film also shot on Stages 8, 9, and 15.

The Kidnapping of the President (Crown International 1980): Hal Holbrook is the president.

The Hand (Orion 1981): assorted scenes. "We were shooting on this stage where DeMille had worked, and Welles, and it was very exciting,"[12] director Oliver Stone remembered, decades later, about filming on the stage.

The A-Team (TV 1983–87): occasional "effects shots."

Mike's Murder (Warner Bros. 1984): assorted scenes.

The Natural (TriStar 1984): alleyway and train scenes; one of Robert Redford's quintessential roles.

Perfect (Columbia 1985): swing sets built here and on Stage 9.

Space Camp (20th Century Fox 1986): space shuttle interiors; other scenes shot on Stages 10, 12, 14, and 15.

Planes, Trains & Automobiles (Paramount 1987): assorted scenes.

V.I. Warshawski (Disney 1991): assorted scenes.

Bill and Ted's Bogus Journey (Orion 1991): assorted scenes, here and on Stage 15.

Man of the People (TV 1991–92): James Garner brought his easygoing charm to this stage for this short-lived series.

Death Becomes Her (Universal 1992): minimal sequences shot here. Most of the film was shot at the Raleigh Studios in Hollywood.

Terminal Velocity (Disney 1994): assorted airborne interiors, here and on Stage 10.

Crimson Tide (Disney 1995): assorted scenes; also used Stages 4 and 15.

Armageddon (Disney 1998): Most of the space shuttle set scenes were shot here, although the film also utilized Stages 4, 10, 14, and 15.

The Matrix (Warner Bros. 1999): The studio claims that part of this game-changer science fiction film was shot here. Maybe. It is certain that they rented this stage as a gym and for rehearsals.

Galaxy Quest (DreamWorks 2000): amusing *Star Trek* spoof; also shot on Stage 9 and at Warner Hollywood Studios.

House of Sand and Fog (DreamWorks 2003): assorted scenes.

Gigli (Columbia 2003): The infamous Ben Affleck / Jennifer Lopez flop shot here, briefly, building an apartment set with a pool dug into the floor on the stage.

Matchstick Men (Warner Bros. 2003): assorted scenes.

Bewitched (Columbia 2005): assorted scenes; also shot on Stages 2, 8, 9, 11, 12, and 15 and around the lot.

Blades of Glory (Paramount 2007): "Olympic habitat."

Rush Hour 3 (New Line 2007): assorted scenes.

21. Makeup and Dressing Rooms

Dressing rooms on a studio property could either be functional little cubicles like those found here, lined up like shoe boxes along the back wall of the studio, or be the lavish bungalows built for Gloria Swanson and her peers elsewhere on the lot. In some cases, a star was afforded both—the bungalow for relaxing or sulking or loving, and the dressing rooms and nearby makeup rooms for the actual application of what would be photographed with that performer on the stages. Often a portable room was placed right on the soundstage for convenience.

Occasionally the needs of the star were dictated not by the actual needs of that star, but rather by that star's ego. A 1945 memo on the subject found the usually

The makeup department and many of the star dressing rooms were housed behind these doors, along this open hallway facing Van Buren Place, during the Desilu era. (2015)

less-than-diplomatic Selznick in a conciliatory mood. It seems that Ingrid Bergman had recently demanded that her RKO (Gower) dressing room follow her to the Selznick lot. Selznick acceded to this demand, but then worried on the record about other stars demanding the same accommodations. He noted that it was "very nice" of another star, Jennifer Jones, to not request any changes in *her* dressing room, then mentioned that "as Miss. Jones' stature as a star grows, and as she sees Miss. Bergman's dressing room (the one borrowed from RKO), and those of other stars, it is obvious that she is going to wish for one although it is almost certain she will say nothing." Selznick's fawning tone regarding Jennifer Jones was almost certainly a result of his personal involvement with the actress, rather than his keen instincts as a producer.

Next door to the dressing rooms, on the western side of the building, was, for many years, the studio makeup department. In 1951 a survey of that department's assets included three long makeup tables with mirrors and lights (value appraised at $64.14 each), twenty-three wig blocks, a hydraulic barber chair, a wig oven, three Halliwell brand "head dryers," and an "Oster massagette"—which, it turns out, was a scalp vibrator. This ruthless inventory also includes a dark brown hairpiece,

Vivien Leigh looks somewhat unhappy with the makeup department's best efforts for *Gone with the Wind.* (1939)

manufactured for the actress Alida Valli, who had done *The Paradine Case* (SRO 1947) and *The Third Man* (SRO 1950) for Selznick and *The Miracle of the Bells* (RKO 1948) for RKO, and apparently happily left it behind on the shelf when she eventually returned home to Italy.

By the time of our survey, makeup was one of the many departments that had been contracted to outside contractors by Desi Arnaz, and these rooms were being used mostly for the storage and use of hair and makeup appliances.

22. Water Tower

A studio water tower has always been a symbol of the might of corporate Hollywood. Even today, most of the surviving studios maintain their water towers, even if those water towers no longer contain any water. The iconography of a water tower, and of the long shadow it casts over the rest of the studio, apparently is more important than its practical usage.

The 100,000-gallon-capacity studio water tower on the lot in Culver City was built in 1927. There don't seem to be any photographs of the water tower with DeMille's name on it, although he certainly did have it painted there. RKO and Selznick and Desilu did the same thing.

The water tower received its most exciting role in real life on October 17, 1964, when one Richard Ulroy, disgruntled and depressed because he could not find employment as a film editor, climbed to the top of the tank, apparently for the purpose of committing suicide. Culver City firefighters summoned to the lot spent several hours trying to convince Ulroy to come down. He finally did, after representatives of the local motion pictures editors' union showed up to listen to his grievances.

Grant Tinker, a television executive from a different medium and a different generation than his forbearers, had the tank taken down in the late 1980s when he was informed that, inexplicably, the city was supplying the studio's water and the water tower, by then admittedly rusted and decrepit, served no practical purpose and hadn't, in fact, for years. Employees remember the tank, as well as an old incinerator that stood next to it, being swarmed by "rednecks with butane torches," who quickly cut the once-mighty giant into scrap—and then carted it away to God knows where.

The studio water tower, seen here during the Desilu era, has always carried the name of whomever was currently in charge. (1964)

23. Stage 10 (79 × 39 feet, 30 feet high, 3,081 square feet)

From the outside Stage 10 looks more like a large house, with a triangular roof and single-story "porch" section, than a place where movies are made. The smallest stage on the lot, at scarcely more than 3,000 square feet, it has, in fact, not been used often for actual production. There are very few mentions of the stage in any context in the records of the studio's various owners. Perhaps it was utilized for still photography, sound recording, or inserts, or for second-unit shots, some of which might not have been noted in the production paperwork. It is also possible that the building was used for assembling or disassembling sets to be stored in the nearby Scene Dock.

The structure itself has an interesting history. What would become Stage 10 was constructed in 1919, near Stage 2, and would be moved across the lot circa 1929, for reasons as mysterious as the rest of the space's history.

The tiny Stage 10 never really looked like a soundstage at all, although it was, and is. (1991)

Occasionally the stage is still utilized, even in modern Hollywood. For Tom Hanks's *That Thing You Do!* (20th Century Fox 1996), for example, the production built, shot, and struck sets inside the building—all over the course of three undoubtedly hectic days in 1995.

Among the few films that admitted to being shot, partially, on Stage 10 are:

Rebound (Pathé 1931): retakes of the "Truesdale" and "Crawford" homes.

Prestige (RKO 1931): also shot on Stages 3, 4, and 12 and on 40 Acres.

Lonely Wives (RKO 1931): assorted scenes; also shot on Stages 4 and 7.

A Woman Commands (RKO 1932): also shot all of over the lot and backlot.

Carnival Boat (RKO 1932): also shot on Stage 12.

The Young in Heart (United Artists 1938): Dubbing and train miniatures were done here. Larger scenes shot on Stages 3, 4, 11, 12, and 14.

Gone with the Wind (MGM 1939): August 24, 1939, "sound effects and wild lines only."

Portrait of Jennie (SRO 1948): On December 4, 1948, Technicolor cameras shot the scene of the title portrait (actually painted by artist Robert Brackman) being admired in an art gallery by students, including one played by a pre-stardom Anne Francis, who wonder if such a vision could ever have been real. "She was real to him," one of the girls exclaims, "or it couldn't look so alive."

FRONT LOT: SETS, SETTINGS, AND SET PIECES

24. Scene Dock

The old Scene Dock building, originally constructed in 1933, offered 20,000 square feet of real estate where set walls and oversize props and assorted ephemera from the movies had long been stored. Next door was a blacksmith shop, built in June 1931. That little building, if not the blacksmith inside, is still there. Interestingly, the roof of both buildings played the top of a garage in *The Racket* (RKO 1951). "If you can't rent it from us we will build it" trumpeted an undated (circa 1980s) ad slick for the department, which also noted that "most major credit cards are accepted."

Unfortunately, as Hollywood gentrified, large plywood walls painted

Half museum, half graveyard, the storied Scene Dock at Culver Studios. (1991)

This little house, improbably wedged amidst the catacombs of the studio Scene Dock, was originally a blacksmith shop. The very same house, if not the smithy inside, is still there. (2015)

with fake brick and statues of long-dead pharaohs built of chicken wire and plaster became less prized than office space and parking space, although to give credit where credit is due, the various managements of the various regimes in the Mansion kept the department largely intact even after the enormous amount of real estate it occupied could no longer be justified by whatever money it managed to produce.

During the Sony era, when another studio was the primary focus of this gradual gentrification, little was changed, but subsequent managements first chipped away and finally eliminated the Scene Dock and its by-then certainly apprehensive staff entirely. Some of the rentable flats ended up at the Downey Studios in Downey, California, which itself was demolished in 2012. Today, the building is still intact, at the moment. It houses assembly and construction space under the rather catch-all title of "Works Department."

The nadir for the Scene Dock came earlier, however, during the ultimately ill-fated Laird era. Until this time employees in the department had, like some mystical Hollywood branch of the Knights Templar, jealously, and generationally, guarded some set walls that stood in the darkest corners of the department. These pieces, in fact, represented many of the interiors from *Gone with the Wind*. It was said that nearly the entire movie could have been re-created from these walls, so well and so completely were they saved.

Unfortunately, there was one employee at the time who was allegedly the son of one of the owners of the studio. One day the entire department rather unwisely went out for lunch, leaving this heir apparent in charge of the Scene Dock. At the time, these pieces were roped off with yellow tape from the rest of the flats and a "Do Not Rent" sign was affixed to them, which this employee took to mean that these sets could, in fact, be destroyed. He therefore felt justified in climbing aboard a forklift and proceeding to do just that.

The only items salvaged from the resultant carnage was a collection of stained glass window panels—which are actually painted glass, designed to photograph like stained glass on-screen. Allegedly, these panels still exist, leaning against a wall somewhere in a corner of this very building.

One of the two *Gone with the Wind* stained glass windows that survive today, thanks to the diligence of several generations of Scene Dock employees. In the film, the middle piece of this window was destroyed by a Yankee cannon ball. (Photo courtesy of John Bertram.)

You can also see these items in the movie, in several scenes. One of them played a window in Rhett Butler's gaudy estate. Others represented the church windows, which are visible in several scenes, including as the backdrop for the sequence where a dying soldier talks about the "wild plum tree that comes to flower in the springtime."

CHAPTER 4

40 Acres: Hollywood Is a Facade

> Do you mean to tell me, Katie Scarlett O'Hara, that Tara, that land doesn't mean anything to you? Why, land is the only thing in the world worth workin' for, worth dyin' for, because it's the only thing that lasts.
>
> —GERALD O'HARA, GONE WITH THE WIND

1. Ince Gate

This was the main entrance into 40 Acres throughout much of its life, named after a man who never lived to use it. A smaller gate was on Higuera Street, east of Schaefer Street. At either gate, there was never exactly a red carpet furled down the driveway for film stars, or film fans. In fact, the rather baroque words "Dogs on Duty" were somewhat more prominent than even the name of the studio (whatever that name was at the time).

The truth is, 40 Acres never *felt* like a Hollywood studio—certainly not in the way that MGM did. MGM, which remember was just a few quiet blocks away, constantly, day and night, had fans waiting outside *its* gates, fans who would shriek at every limousine that pulled up to its ornate white colonnade. In contrast, 40 Acres was in the middle of a suburb! The tract houses and warehouses and industrial parks in this neighborhood were where people dreamed about the movies, and about the movie stars who appeared in them, certainly not, absolutely not, where those stars *made* those movies, and those dreams, for the rest of us.

Both gates, throughout their uneventful lives, were manned by bored and underpaid security personnel, who probably dreamed about Hollywood along with

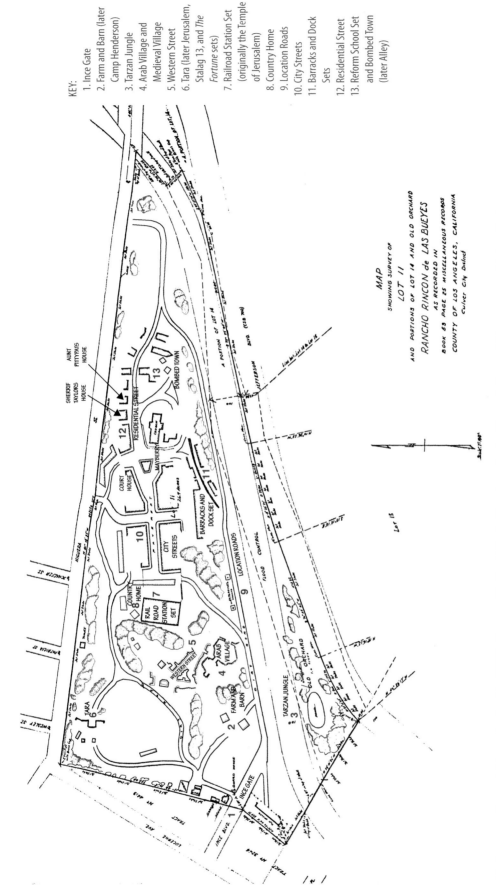

KEY:
1. Ince Gate
2. Farm and Barn (later Camp Henderson)
3. Tarzan Jungle
4. Arab Village and Medieval Village
5. Western Street
6. Tara (later Jerusalem, Stalag 13, and *The Fortune* sets)
7. Railroad Station Set (originally the Temple of Jerusalem)
8. Country Home
9. Location Roads
10. City Streets
11. Barracks and Dock Sets
12. Residential Street
13. Reform School Set and Bombed Town (later Alley)

40 Acres as it would have looked in 1959. *Gone with the Wind*'s Tara is noted, although it would be removed later that year, as are several already standing locations from *The Andy Griffith Show*, which would not go into production until 1960. (Map by Robert Lane, adapted from a 1956 survey.)

Front lot and backlot. (1938).

the rest of us. Even during the flickering twilight of the studio's life, Rex McGee, a film student at USC, anxious to explore 40 Acres, remembered that Leo Pepin, the operations manager at the studio at the time, had to dutifully call him in a pass so that that a lonely security guard, at that lonely gate, could check his name off a lonely list, upon which his was the only name, and let him explore a ghost town where no one else was working.

That job at the Ince Gate did have its perks. The gate, and that guard shack, occasionally appeared on camera as a set—as a guard shack playing a guard shack. A 1966 episode of *Batman* (TV 1966–68), "Holy Rat Race," for example, used the gate, as did *Gomer Pyle, U.S.M.C.* (TV 1964–69), where the gate and the guard shack, if not the guard inside it, was used several times as the entrance to "Fort Henderson." The actual guard on these days might have, with luck, even been able to meet Jim Nabors.

For many years a caretaker's shed stood next to the Ince gate. This unnamed caretaker, whose name has been lost to history, would have been responsible for the operations, maintenance, and security of the entire property. The stories he could have told . . .

40 ACRES: HOLLYWOOD IS A FACADE

Marc Wanamaker, contributor to the current volume, poses outside the Ince Gate. (1969)

2. Farm and Barn (later Camp Henderson)

Throughout its existence, it should be noted that there were an awful lot of farmhouse sets at 40 Acres. This can be explained, partially, by the first hundred-plus years of America's own existence.

During this period, the United States largely was an agrarian society. Most of its population lived on, or around, farms. Even as the swelling population moved into the cities, the shared memories of this common background continued to be an influence on a population that worked in offices and wore gray flannel suits, yet laughed at Ma and Pa Kettle and the Real McCoys and the Beverly Hillbillies, and in their dreams rode the range with William S. Hart and John Wayne.

The Real McCoys (TV 1957–63) would seem to be a reflection of this tradition, yet few today, of those who remember the series at all, recall that it was about a West Virginia family that relocated in the San Fernando Valley in Los Angeles, a farm family integrating itself into the suburbs. The show actually reflected the ongoing

The Farm and Barn set as seen in *The Real McCoys*. (1957)

urbanization of the United States, just as *The Andy Griffith Show* (TV 1960–68) would gently protest that urbanization, and *Gomer Pyle, U.S.M.C.* (TV 1964–69) would blindly accept it. All three of these series, it should be noted, would shoot here, in exactly the same place.

This farm was a one-story structure with a porch and adjacent barn, both of which also showed up in episodes of *U.S. Marshal* (TV 1958–60). More interesting, perhaps, than the set itself was the amount of real estate the set occupied, with actual roads and fields and trees filling in the background and giving the location a most impressive sense of reality most backlots could only envy.

The very familiar "Goober's Service Station" set near the end of its life. (1970, photo courtesy of Rex McGee.)

That spaciousness also made the set vulnerable. In 1964 the McCoy house and barn were removed and *Gomer Pyle*'s Camp Henderson Quonset huts and barracks were constructed on the same site, although the field and knoll in the background would remain intact. *The Andy Griffith Show* was still on the air at the same time and "Wally's Service Station," known later in the series and in *Mayberry R.F.D.* as "Goober's Service Station," would eventually occupy the site as well. As spacious as the area still was, occasionally those California-set Quonset huts could be glimpsed behind Wally's garage.

Those Quonset huts and that gas station had a life beyond the series they were built, or requisitioned, for. The huts, for example, were seen, gutted and empty, in a 1966 episode of *Batman*, "Holy Rat Race." *The New People* (TV 1969–70) also used the very same huts in a mysterious tropical milieu, foretelling the plot of another series, *Lost* (TV 2004–10), which would confuse viewers decades later. The gas station showed up in an episode of *Gomer Pyle* (as a diner) and probably elsewhere as well.

Yet another apparently entirely different farm, slightly west of this set, appeared, according to a 1936 article in *Stage* magazine, in *The Animal Kingdom* (RKO 1932) as well as "hundreds of other productions."[1] A viewing of *The Animal Kingdom* reveals nothing similar to this structure, however, so we will have to take the anonymous author's word about the rest as well. A 1940 studio map identifies this farm as the "Country Doctor" set, which, as film historian Kipp Teague points out, was the name of a 1927 DeMille (produced) picture that, unfortunately, exists today only in fragments, so this one will have to remain a mystery.

Another farmhouse as well as an elaborate French village, probably built for *The Unknown Soldier* (PDC 1926) or *The Fighting Eagle* (Pathé 1927), once stood on or near this site, which stretched across both sides of Ballona Creek, the two sections connected by a stone bridge. The bridge was apparently first seen in *The Leatherneck* (Pathé 1929). *All Quiet on the Western Front* (Universal 1930) adapted

The Fort Henderson Quonset huts as repurposed in the short-lived series *The New People*. (1970, photo courtesy of Rex McGee.)

120 HOLLYWOOD'S LOST BACKLOT

Part of the original French village set that spanned Ballona Creek and was notably used in *All Quiet on the Western Front.* (1931)

these sets at 40 Acres for their Oscar-winning antiwar masterpiece. Jefferson Boulevard, where the church was blown up in the movie, was at the time a small dirt road and the southern border of the lot. Ballona Creek was also used for swimming scenes before the battle began. For these scenes the film company dammed up the creek to create enough slow-moving water to swim in.

After Universal finished using the lot, those French village sets were reused for several years and, of course, modified for other films, such as *Westward Passage* (Pathé 1932), *The Count of Monte Cristo* (United Artists 1934), and *The Three Musketeers* (RKO 1935). Sadly, that very pictorial stone bridge, so prominent in *All Quiet*, had to be eliminated in the mid-1930s when the Army Corp of Engineers channelized the creek and lined it with concrete. A section of the Jefferson Boulevard side of the property was later converted into a jungle set and used for many years (see "Tarzan Jungle" below).

3. Tarzan Jungle

In the early 1930s, MGM acquired the rights to Edgar Rice Burroughs's Tarzan character and, starting in 1932 with *Tarzan the Ape Man*, made six high-quality features on the subject (from 1918 through the 1930s, eleven Tarzan films were made by companies other than MGM). These MGM productions starred Johnny Weissmuller as the ape-man and Maureen O'Sullivan as Jane. After *Tarzan's New York Adventure* (1942), however, MGM lost interest in the series and the rights to the character reverted to producer Sol Lesser, who had earlier produced the low-budget *Tarzan the Fearless* (Principal Distributing 1933) and had been making competing Southern California–filmed jungle epics ever since. Lesser would continue to make jungle films, some with different stars, and different studios, and different jungles, until 1958.

Lesser also had a long, if pockmarked, history with RKO; he had made *Return of Chandu* (Principal Distributing 1934) at 40 Acres, for example. Lesser and Tarzan himself, Johnny Weissmuller, obviously realized that a distribution deal with a major like RKO would give them access to both production and distribution avenues, which they could never surmount as independents. So, in 1943 it was announced that *Tarzan Triumphs* would be shot at 40 Acres and distributed by RKO. The only casualty in this deal was O'Sullivan, who was still under contract at MGM and from all accounts was not unhappy about leaving her vine-swinging days behind her.

The six Tarzan epics that would follow would themselves be followed by several more Tarzan features with younger actors, most of them shot in a narrow jungle wedge on the other side of Ballona Creek. Inadvertent though it might have been, this close proximity to the creek quickly proved to be a most invaluable asset for productions. In addition to providing water and water-based jungle locations, Ballona Creek offered a place from which to look into the jungle—which has always proven to be extremely difficult for filmmakers forced to try to photograph an actual jungle from *inside* that jungle.

Make no mistake, the Tarzan Jungle was indeed *very* small, yet it never felt that way on-screen. As small as it was, the set decorators and greensmen somehow always found room to stock it with a rocky ledge, a pond, some densely wooded areas for Tarzan to run and swing through, and part of Tarzan's treehouse as well, which copied the MGM version and was completed by a matte painting. The Baldwin Hills, in the background, also were often conscripted, quite convincingly, to play the verdant hills of Africa.

Visual evidence would seemingly suggest that this jungle was first implemented in response to the early 1930s craze for jungle epics, which had been partially touched off by MGM's *Tarzans* and which RKO had contributed to with *Bird of Paradise* (1932), *The Most Dangerous Game* (1932), and *King Kong* (1933). In fact, the earliest version of the jungle seems to have originated for *Bird of Paradise*. Apparently, the jungle was purpose-built—on the southern side of Ballona Creek, near where

The RKO Tarzan Jungle included approximately half of Tarzan's treehouse. The other half, as well as some of the vegetation, was a clever matte painting. (1951, top photo courtesy of Joseph Musso and Walter O'Connor; bottom photo courtesy of Joseph Musso.)

some of those *All Quiet on the Western Front* sets had stood, on real estate identified on Culver City maps and deeds as the "old orchard"—for this movie, or for some other rainforest odyssey during this era.

It is possible, however, that some familiar prop trees and rocks migrated over from other films and other locations and ended up filling out Tarzan's jungle as well. Or is it possible that trying visually to match trees and rocks in back of Johnny Weissmuller's acrobatics makes one look for patterns that are not, in fact, there?

Additionally shot in the Tarzan Jungle:

The Son of Kong (RKO 1933): "Ext. Temple ledge and pool—edge of Jungle" shot nonspecifically at "Pathé."

The Tuttles of Tahiti (RKO 1942): assorted scenes.

Tarzan's Desert Mystery (RKO 1943): assorted scenes.

Tarzan and the Amazons (RKO 1945): also shot some scenes at the LA Arboretum in Arcadia.

First Yank into Tokyo (RKO 1945): supply depot and hospital exteriors.

Tarzan and the Leopard Woman (RKO 1946): also shot some scenes at the LA Arboretum in Arcadia

Tarzan and the Huntress (RKO 1947): again, also shot some scenes at the LA Arboretum in Arcadia.

Tarzan and the Mermaids (RKO 1948): shot in Acapulco, mostly.

Tarzan's Magic Fountain (RKO 1949): Lex Barker here replaces Johnny Weissmuller as the ape-man.

Tarzan and the Slave Girl (RKO 1950): Lex Barker again.

Tarzan's Peril (RKO 1951): Lex Barker again. RKO injected some new energy into their aging franchise here by allegedly shooting some of the film in Africa, although 40 Acres is in there too.

Tarzan's Savage Fury (RKO 1952): yes, Lex Barker.

Tarzan and the She-Devil (RKO 1953): Lex Barker and Monique van Vooren play the title roles.

Tarzan's Hidden Jungle (RKO 1955): Gordon Scott's first Tarzan film, and RKO's last. Ironically the series, and Scott, would move back to MGM for the next film, *Tarzan and the Lost Safari*, which would be filmed not on the lot but overseas.

The Adventures of Jim Bowie (TV 1956–58): shot assorted Louisiana bayou scenes here in various episodes.

Around the World in 80 Days (United Artists 1956): Historian / production illustrator Joseph Musso uncovered this set's appearance, doubling as India, in this all-star extravaganza.

The Real McCoys (TV 1957–63): used the set in the 1963 episode "Pals," and probably others.

The Andy Griffith Show (TV 1960–1968): assorted scenes.

Ride Beyond Vengeance (Columbia 1966): assorted scenes.

Surely, there were many more indeed.

4. Arab Village and Medieval Village

One of the most exotic sections on all of 40 Acres was a labyrinthine cluster of "foreign" buildings that could provide, as a 1936 article in *Stage* magazine put it, "English, French, Spanish and Russian Architecture, if the cameraman is clever and the art director not too fussy."[2]

The medieval section on the western side of the set seems to have come first, apparently from *The Night of Love* (UA 1927), a big-budget Samuel Goldwyn project

The Arab Village, center, was a labyrinthine maze of walls and alleys and markets, all with an overtly "foreign" air, which made it ideal for exotic locales. (1963)

that was, incidentally, set in Spain, and which was followed by the Gaelic-set *The Fighting Eagle* (Pathé 1927). Although this section included a castle gate and tower (sometimes called the "Biblical Castle," indicating an even earlier *King of Kings*–era origin), medieval cottages, and cobblestone streets, it was not used often after that.

When the Medieval Village was utilized, it was in low-budget rental projects like the dire Bela Lugosi serial *Return of Chandu* (Principal Distributing 1934) and a Boris Karloff companion, *The Black Room* (Columbia 1935). That more prestigious rental project, *All Quiet on the Western Front* (Universal 1930), also probably used part of the set, although most of the 40 Acres sequences were shot, as noted, slightly south of this village in sets partially constructed (and then partially destroyed) for that film. Eventually most of the Medieval Village was cannibalized and adopted and integrated into the nearby Arab Village, which it already shared some transitional structures with.

The Arab Village's ethnic origins, despite its assertive name, are just as tangled as the Medieval Village's. In fact, the set was often referred to, internally and inconstantly, as "Arabian Court," "Chinese Street," "Hindu Street," "Moorish Street," or as the "Latin Quarter," depending on what it needed to impersonate at the moment. It was apparently constructed, at a cost of $18,227.44, for *Turkish Delight* (PDC 1927), but that movie's geography, despite the title and like the set itself, was somewhat murky. *The Forbidden Woman* (Pathé 1927) and *Fighting Love* (PDC 1927), both of which were set in North Africa, and the Holy Land–set *King of Kings* (again DeMille, again 1927) followed, and deepened the confusion.

Selznick's *The Garden of Allah* (United Artists 1936) was also set in North Africa and probably cemented the "Arab Village" moniker onto the set, if any one name could ever be used to describe the location. This is one of those films where Marlene Dietrich effervescently glides across the shifting desert dunes, in high heels. Those dunes, however, were neither in North Africa nor on 40 Acres but in Yuma,

The Medieval Village's European ambiance (1936, left) eventually was folded into the even more exotic Arabian set (which is *almost* visible behind that castle in 1927).

Arizona, exactly where part of *Star Wars: Episode VI—Return of the Jedi* (1983 20th Century Fox) would be shot, seemingly a thousand years later.

The set was often—as noted, and for undocumented reasons probably related to its exotic milieu—used as a place for lunches, parties, and press conferences at 40 Acres.

In December 1975 the Arab Village, now inexplicably referred to as the "Spanish Compound," burned to the ground in yet another mysterious fire. Gaylon Botten, a spokesperson for Culver City Studios at the time, flippantly told the press that there was "no dollar loss associated with the property. We planned to demolish it in January. It had been previously vandalized and virtually destroyed, and part of it had burned in an earlier fire." Botten added glibly, "We're not using the backlot at all. We're hoping to sell it for an industrial project."[3]

Additionally shot at the Arab Village and Medieval Village:

Blondes by Proxy (RKO 1931): The Arab Village played a "Moorish street and garden" for this "Traveling Man" short subject.

A Woman Commands (RKO 1932): assorted scenes.

Prestige (RKO 1932): as Indochina, or something.

Roar of the Dragon (RKO 1932): Manchuria is here played by "Chinese Street."

The Monkey's Paw (RKO 1932): "Hindu Street," as it was then called, played an unidentified village in an unidentified Eastern European principality.

Secrets of the French Police (RKO 1932): "Chateau Streets."

King Kong (RKO 1933): Nothing in the film matches anything on the set, yet the daily production reports attest that *something* for this legendary production did, in fact, shoot on "Hindu Street."

The Last Days of Pompeii (RKO 1935): as the "slave village."

The Young in Heart (United Artists 1938): As in the later *Rebecca* (United Artists 1940), Monte Carlo was the location.

Intermezzo: A Love Story (United Artists 1939): Both the Arab Village and the Medieval Village, in one of their more ethnically diverse roles, here costarred with Ingrid Bergman as Stockholm.

Rebecca (United Artists 1940): Like Ingrid Bergman, Alfred Hitchcock made his American debut in the Arab Village, shooting the marriage scene between Joan Fontaine and Laurence Olivier, which according to the script occurred in Monte Carlo.

Tarzan Triumphs (RKO 1943): The Arab Village is now the exotic "City of Palandra." This would be the first of many Tarzan visits to the set.

Asia was not beyond the scope of the Arab Village's range, as demonstrated by this unidentified but exotic 1944 set still.

Tarzan's Desert Mystery (RKO 1943): Tarzan swings through for a repeat visit. The setting was, at least here, somewhat Arabic.

I Walked with a Zombie (RRO 1943): the "St. Sebastian Street Café" somewhere in the Caribbean, as portrayed, of course, by the Arab Village.

Rookies in Burma (RKO 1943): RKO's Abbott and Costello, (Allen) Carney and (Wally) Brown, visit what is here known as "Burmese Street."

Behind the Rising Sun (RKO 1943): "Chinese Village."

Around the World (RKO 1943): "Street of Bazaars."

The Falcon in Mexico (RKO 1944): Mexico City.

Tarzan and the Amazons (RKO 1945): Tarzan visits the "City of the Amazons."

The Arab Village in its later years. (circa 1960s)

First Yank into Tokyo (RKO 1945): The Arab Village is now the "Yamuri compound."

China Sky (RKO 1945): a "village and compound" in China, as portrayed by a village and compound on 40 Acres. What was left of the medieval section was also utilized.

Tarzan and the Leopard Woman (RKO 1946): assorted scenes.

Tarzan and the Huntress (RKO 1947): assorted scenes.

Tarzan's Magic Fountain (RKO 1949): assorted scenes.

Macao (RKO 1952, filmed in 1950): "Rua da Felicidade."

Viva Zapata! (20th Century Fox 1952): Part of the climax, where Marlon Brando's Zapata is assassinated, was shot here, with the set lavishly redressed as old Mexico.

One Minute to Zero (RKO 1952): Robert Mitchum visits South Korea, and the Arab Village.

Tarzan and the She-Devil (RKO 1953): the end of an era, as this was the last Tarzan film shot on the set.

The Conqueror (RKO 1956): The Arab Village here is seen as the "Urga gates and marketplace" for Howard Hughes's expensive and misguided biography of Genghis Khan—played with prosthetically "slanted" eyes and a Fu Manchu mustache by John Wayne.

Much of *The Conqueror* was shot on location in St. George, Utah, in 1953. Unfortunately, in nearby Yucca Flats, Nevada, the US military had just finished detonating eleven nuclear bombs as part of its atomic testing program. Radioactive fallout from these tests drifted into Snow Canyon, Utah, and then up onto the set of *The Conqueror,* which RKO was using as a stand-in for ancient Mongolia. Unaware of the potential danger, the cast staged barbarian battles and torrid seduction scenes between Wayne and costar Susan Hayward amid those swirling clouds of dust, and whatever was inside that dust.

Tragically, when the location shoot was completed and the crew returned to Hollywood, some sixty tons of this red, radioactive debris was shipped back with them so that the footage shot in the stages and on the backlot would match that from the location. Worse yet, when the film wrapped, the crew trucked what was left of this radioactive dust up into the Baldwin Hills and dumped it in areas that today contain Culver City suburbs.

The results of this nuclear nightmare were tragic indeed. Of the roughly 220 people who worked on *The Conqueror,* 91 had contracted cancer as of the early 1980s and 46 ultimately died of it, including stars Wayne, Hayward, and Agnes Moorehead and director Dick Powell. Another one of the film's stars, Pedro Armendáriz, who had just finished his role in *From Russia with Love* (United Artists 1963), shot himself through the heart when given his own cancer diagnosis.

In 1980 Dr. Robert C. Pendleton, a director of radiological health at the University of Utah, studied what had happened on this set and stated that "with these numbers, this case could qualify as an epidemic. The connection between fallout radiation and cancer in individual cases has been practically impossible to prove conclusively. But in a group this size you'd expect only 30-some cancers to develop. With 91, I think the tie-in to their exposure on the set of *The Conqueror* would hold up even in a court of law."[4]

Although threatened lawsuits against the United States government never came about, most of the cast and crew have continued to report cases of cancer to the present day (not to mention the countless cases that are likely unreported or unrealized), often now among children and grandchildren of those who survived *The Conqueror,* and those who did not.

Howard Hughes was reportedly devastated by the reputed links between his film and the misfortunes of so many of its cast. He earlier had withdrawn *The Conqueror* from circulation, although Wayne's apparently undimmable box-office clout had made it an unlikely box-office success in 1956. At the end of his life, when no one else in the world could watch his picture, Hughes screened it hundreds and hundreds of times, reciting the movie's dialogue alone in his darkened penthouse and quietly tabulating its many victims.

Son of Sinbad (RKO 1955): assorted Arabian escapades.

The Texan (TV 1958–60): The Arab Village here is, in at least one episode, a Spanish village.

The Greatest Story Ever Told (United Artists 1965): Was this only the second time Jesus came to 40 Acres? The village, along with many purpose-built additions, once again played Jerusalem.

Harum Scarum (MGM 1965): Elvis Presley visits the Arab Village for "exterior, Market" scenes.

Star Trek (TV 1966–69): In the 1960s the set started appearing with semiregularity on television. It appeared in the original pilot of *Star Trek* (TV 1966–69), "The Cage," which was rejected by the network, although the footage did show up in a later two-part episode, "The Menagerie."

Batman (TV 1966–68): occasional assorted scenes.

Hogan's Heroes (TV 1965–71): assorted (German!) locations.

Mission Impossible (TV 1966–73): assorted scenes, as various exotic locales.

The Arab Village in twilight. (1970, photo courtesy of Rex McGee.)

5. Western Street

It is perhaps surprising on a gut level that this most quintessential of studios, for many years, had no permanent Western street on its backlot, although a short-lived version seems to have once stood south of the Ince Gate. Ince and DeMille both specialized in Westerns, but 40 Acres came after the former had departed, and DeMille's big-scale frontier sagas were all produced after he had returned to Paramount. RKO did have a large Western set, built for *Cimarron* (RKO 1931), but it stood at their Encino Ranch. And David O. Selznick did not particularly like the genre. Once when asked how far the Western had come, the producer replied "from Wyoming to Arizona—and back."[5] Selznick's one big Western, *Duel in the Sun* (SRO 1946), had rented its Paradise Flats "town" sets from MGM.

So, it was left to that unlikely producer Desi Arnaz to bring the Old West to 40 Acres.

The Western Street was somewhat Y-shaped and was geographically situated between the Arab Village and the Railroad Station Set. Built primarily for *The Texan* (TV 1958–60), there is some evidence that an embryonic version of the location had been used earlier, for *The First Traveling Saleslady* (RKO 1956). *Traveling Saleslady*, whether it introduced the set or not, did feature, in major supporting roles, James Arness and Clint Eastwood vying for stars Ginger Rogers and Carol Channing. Arness and Eastwood, as we know, would subsequently ride the range, respectively, in television and in feature films, for decades.

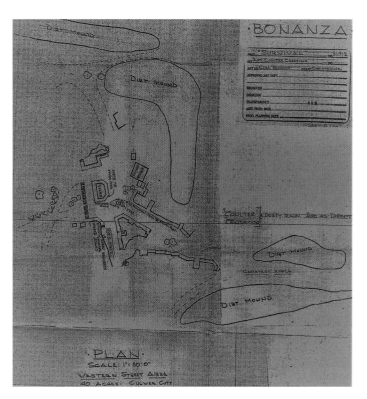

The 40 Acres Western Street layout as adapted in 1967 for *Bonanza*.

The Western Street was assembled, frankly, on the cheap, because Desi intended to use it, and rent it out, primarily for television. The year before, Warner Bros. had built a similar set on their Burbank lot, which they initially called the "TV Western Set." The construction of both sets was a response to an unprecedented boom in the production of Westerns for the small screen. In 1959, the year after the 40 Acres set made its dusty debut, there were a remarkable *twenty-six* Westerns crowding the prime-time schedule.

Unfortunately, the set was completed at the top of this trend. Westerns would continue to be popular for the next several years, but as the craze trailed off, more and more locations to shoot these Westerns became available. The 40 Acres version

132   HOLLYWOOD'S LOST BACKLOT

The Western Street seemingly looked into Texas on one side (top) and Arabia on the other (bottom). (1969)

The Western Street in 1974 apparently boasted real tumbleweeds.

had to compete with sets built for feature films—which were both larger and more detailed—and with movie ranches, properties usually north of Hollywood, which offered producers more visual variety and natural backdrops. Eventually the set ended up earning its keep by portraying what could be described as almost a parody of what a Western street was supposed to look and feel like, as contemporary-set series like *The Dick Van Dyke Show* and *My Three Sons* used it to spoof Western conventions or to place their twentieth-century characters in a nineteenth-century-style setting.

Additionally shot on the Western Street:

The Adventures of Ozzie and Harriet (TV 1952–66): as a ghost town, in the 1965 episode "The Ghost Town."

The Adventures of Jim Bowie (TV 1956–58): occasional.

The Californians (TV 1957–60): The Western Street made occasional appearances in this California Gold Rush–set series.

Yancy Derringer (TV 1958–59): occasional.

U.S. Marshal (TV 1958–60): A spin-off of Desilu's earlier *The Sheriff of Cochise* (TV 1956–59), this contemporary Western was, unlike its predecessor, which had many actual Arizona locations, mostly shot in California, both here and at the Iverson Movie Ranch in northern Los Angeles County, although the episodes reveal a surprising amount of location work on display as well. The on-screen title for this series is actually "United States Marshal."

Bat Masterson (TV 1958–61): used the set in the 1959 episode "Dead Men Don't Pay Debts," possibly others.

Pony Express (TV 1959–60): occasional.

The Untouchables (TV 1959–63): A famous two-part episode, "The Big Train" (1961), cast Western Street as the (fictional) Cloverville, California, where Eliot Ness and his men foil an attempt to free Al Capone, who is being transported to Alcatraz. The episode highlighted the rarely seen railroad tracks, which forked on the way out of town.

Bonanza (TV 1959–73): One of the most famous, and venerable, of all TV Westerns shot only five episodes in Culver City: "False Witness" (1967), "Catch as Catch Can" (1968), "My Friend, My Enemy" (1968), "The Clarion" (1968), and "The Running Man" (1969), the last of which staged a dramatic burning house sequence on the set.

My Three Sons (TV 1960–72): seen in one episode, "The Horseless Saddle" (1961). Note that the signage on the flats still says "Cloverville," from that recent visit by Eliot Ness and his Untouchables.

The Dick Van Dyke Show (1961–66): A 1966 episode, "The Gunslinger," shot an amusing dream sequence here, one of the few actual exteriors ever utilized in that classic series.

Hogan's Heroes (TV 1965–71): used the barn on the edge of the set at least once.

Ride Beyond Vengeance (Columbia 1966): as Coldiron, Texas.

Batman (TV 1966–68): "Come Back, Shame" a 1966 episode, with Cliff Robertson as that episode's not-so-bright supervillain.

Rango (TV 1967): Tim Conway played the title role in this Western comedy. The episode "If You Can't Take It with You, Don't Go," possibly others, shot here.

The Guns of Will Sonnett (TV 1967–69): Walter Brennan (*The Real McCoys*) returned to 40 Acres, occasionally, for this end-of-the-trail Western.

Lancer (TV 1968–70): as California's San Joaquin Valley.

Heading north past the Western Street, a visitor, at least a visitor in the late 1940s, would have next seen a craggy rock wall, probably constructed out of gunite, for *Mighty Joe Young* (RKO 1949), although the production paperwork actually referred to the location as an "open space and rocky ledge." Whatever it was called, the set consisted of a wall of artificial boulders that were used in a memorable scene in that memorable movie.

Mighty Joe Young used these artificial rocks for some exciting cowboy-gorilla roping action. Producer Merian C. Cooper's empty chair takes it all in. (1949, photo courtesy of Joseph Musso and Walter O'Connor.)

The stately Manderley from *Rebecca* (United Artists 1940)—or at least the doorway and surrounding walls of Manderley—also stood nearby, just south of another stately mansion . . .

6. Tara (later Jerusalem, Stalag 13, and *The Fortune* sets)

> What would I want with a farm?
> —Rhett Butler, Gone with the Wind

Tara is, of course, the name of the plantation house seen in *Gone with the Wind*, which also happens to be the most famous movie set of all time.

It is also, perhaps inevitably, a metaphor. Surprisingly, even today, Tara is widely regarded as something of an allegory, a totem symbolizing a lost and nearly mythical Hollywood that probably never existed, and so can never be re-created, never be recaptured.

The truth is that this simply is not the truth. The symbol might be gone with the wind, but alone amongst all of the sets that have ever stood on 40 Acres, Tara is

The ravages of time. Tara as it looked on the backlot in 1939 and in 1959.

the only one that, almost impossibly, somehow still exists—more than two thousand miles from Culver City.

In Margaret Mitchell's book, the O'Hara family home is described as ugly and ill-conceived, and Mitchell implored Selznick to leave it that way. "I besought them to please leave Tara ugly, sprawling, columnless, and they agreed. I imagine, however, that when it comes to Twelve Oaks, they will put columns all around the house and make it as large as our new city auditorium," she is quoted as saying.[6]

Mitchell was right about what Hollywood would do to Twelve Oaks, the neighboring plantation, which was little-seen in the film and was created for the movie largely using lots of white columns, elaborate matte paintings, and vast interior sets. In fact, regarding Twelve Oaks, when she saw the movie, Mitchell commented privately that "I did not know whether to laugh or to throw up at the *two* staircases."[7]

But Selznick's people misled the author about Tara. The building the producer's art department came up with would, of course, have white columns, 21 feet

Scarlett and Tara, but note the cables connecting the branches of the two "oak" trees. (1939)

tall, although they were made of whitewashed brick and were square, rather than rounded. Tara would, at least, ultimately only have *one* staircase.

Actually, it is telling that Tara was built on the backlot at all, as it actually gets little more screen time in the movie than Twelve Oaks. Clark Gable, the film's leading man, appeared on-camera with the Tara facade for a grand total of 37 seconds. Tara itself, the backlot set, appeared on-screen for a total of 8 minutes and 44 seconds—out of a 3-hour-and-58-minute movie.

Yet there is something to be said for the physicality of actual sets, and those sets' effect on both actors and audiences. Selznick, who had (executive) produced *King Kong*, understood special effects, understood what could be portrayed by a miniature or by a matte painting or by a painted backdrop. For example, for his very next film *Rebecca* (United Artists 1940), he would build that film's estate, Manderley, in miniature on a stage and in pieces on 40 Acres. But he also understood that *not* building Tara or building only part of it—a doorway or a column on a soundstage, for example—would have been cheating both his film and his audience. Just as he insisted his actors be costumed down to the Victorian-era underwear the audience would never see, he realized that an actual house, or rather a backlot set portraying an actual house, would have an impact far beyond that house's actual screen

Although Tara's whitewashed columns and the porch floor were real, the rest of the brick seen here was completely artificial, though highly convincing. (1939)

40 ACRES: HOLLYWOOD IS A FACADE   139

time—which today's producers, in our CGI-enriched era, don't seem to realize, or to recognize.

Ultimately, as author Tommy Jones described it in his definitive study of the set, "the entire front and right side of the main house were completed, parts of the left side of the main house, three sides of the kitchen and all sides of the connecting breezeway. Only the front halves of these structures appear to have been roofed, although cornice was installed down the rear slope of the right side of the main house and on both sides of the rear slope of the kitchen roof."[8] Jones estimates that the entire set ran probably 180 feet across the front and stood about 38 feet high to that rooftop.

Only those iconic white columns and the floor of the porch were actually made of brick. Most of the set was built, as were most backlot structures, then as now, of unpainted lumber covered with plywood or the equivalent, with patterned material

David O. Selznick, on the eve of producing *Gone with the Wind*, gazes from the Railroad Station Set and across 40 Acres—and out to Tara. (1939)

All of David O. Selznick's subsequent productions stood in the shadow of Tara—sometimes literally, as here, with Gregory Peck and Jennifer Jones in *Duel in the Sun*. (1946)

representing the desired architectural detailing attached to the outside. Only the suggestion of the interiors was included on the outdoor facade, such as the interior trim created behind the front entrance and the accurate detailing created on some of the interior window jambs.

Jones notes that the set included numerous examples of the high level of craft and attention to detail that went into making Tara "real." For example, the use of "old fashioned wooden glazing strips rather than putty, to secure some of the glass to the windows, the window sash and wooden pegs, rather than nails, on the steps of the breezeway."[9] While visible in the film, few viewers would have noticed these details.

There are several photos of Selznick walking on his backlot before, during, and after the shooting of *Gone with the Wind*. Tara is in the background in many of them. It would remain in that location for the rest of Selznick's life.

Selznick, apparently, even at the time was aware that *Gone with the Wind* would ultimately become some sort of totem for what he, of what Hollywood, was ultimately capable of conceiving and creating. So it is possible that from the beginning he wanted Tara to be a physical representation of what he rapidly came to see as the defining moment of his career.

40 ACRES: HOLLYWOOD IS A FACADE

And he left the set standing for the next twenty years, even though, after late 1939, those white columns obviously made the set too easily recognized to utilize in any other context. He referred to the facade in interviews throughout his life as well. In 1959 he reflected that "nothing in Hollywood is permanent. Once photographed, life here is ended. It is almost symbolic of Hollywood. Tara has no rooms inside. It is just a facade. So much of Hollywood is a facade."[10]

As noted, the Tara facade would prove to be a very hard set to repurpose for other projects. This would have been a death knell for any ordinary set—but Tara, of course, was not an ordinary set. In fact, it survived for decades, sitting atop its little graded knoll on 40 Acres while other, much more utilized sets rose below it, flourished, aged, and finally blew away. Tara itself also aged. Its brick, both real and simulated, eventually dissolved, its paint flaked, and its doors and window shutters could often be heard squeaking drunkenly at the bidding of the Santa Ana winds from even the most distant corners of the lot.

The set did, contrary to published reports, occasionally appear on-camera during this period, although never again in a Selznick picture. *Little Men* (RKO 1941) used the doorway and a couple of the columns as the "State Orphanage" and also, strangely, as a backdrop for its opening credits. *The Devil and Daniel Webster* (RKO 1941) tried to disguise the plantation by adding a second-floor balcony.

The years passed. *Lady Luck* (RKO 1946) and *Banjo* (RKO 1947) staged some scenes in front of the mansion, but *The Adventures of Superman* (TV 1952–58) gave even the sharpest-eyed audiences only a glimpse of the plantation, in a first-season episode, "The Time Machine." Likewise, the set can be seen, briefly, in the background, in the feature film *Verboten!* (RKO 1959) and in at least one episode of *The Untouchables* (TV 1959–63). An episode of *The Texan* (TV 1958–60), specifically 1958's "A Tree for Planting," gave the rotted walls, now covered with a plank fence, more screen time, if less dignity. Finally, *Yancy Derringer* (TV 1958–59) in a 1958 episode, "Memo to a Firing Squad," thought to cast the plantation house as . . . a plantation house.

Representations of Tara, in the form of copycat sets at other studios, actually had longer and richer on-screen lives than the original. Some of these other Hollywood plantations, in point of fact, actually predated Tara, yet were destined to stand in its shadow forever after 1939. Universal Studios, for example, long-boasted a Southern mansion set, actually called the Colonial Mansion, which was built in 1927, and 20th Century Fox also had a plantation set, called the Colonel House, built in 1934, both of which perhaps inspired, and later evoked, Tara.

At MGM's Backlot 2 there also stood a set called the Southern Mansion, which for decades was identified as Tara, or as Twelve Oaks, by studio executives and employees giving tours of the lot to gullible friends. Of course, it wasn't. That MGM set was instead created for *The Toy Wife* (MGM 1938). When the Southern Mansion was destroyed in 1978, a Pasadena couple, seeing pictures of the house being

Looking through these garbage-strewn chain-link fences in 1956, Higuera Street motorists would have been able to see, and perhaps recognize, the skeletal remains of Tara decomposing on the backlot.

bulldozed, were inspired to build a copy of it. This copy has itself been identified in local guidebooks as having been photographed in *Gone with the Wind*, as either Tara or as Twelve Oaks. And on it goes.

One of the most similar movie-made plantations of all was built at Republic Studios in 1949 for *The Fighting Kentuckian*, which was set not in Kentucky at all but in Alabama. It looked so suspiciously like Tara that when the building was repurposed for the TV series *The Big Valley* (1965–69), many viewers assumed that the Barkley family home was in fact played by the original *Gone with the Wind* set, even though that TV show was not based in the Deep South at all, but in California's San Joaquin Valley.

Other "appearances" of Tara in corrupt popular culture include the legendary 1976 "Went with the Wind" sketch on *The Carol Burnett Show* (TV 1967–78), which utilized a single stock shot of a real plantation, acquired from some dubious source. *Moviola: The Scarlett O'Hara War* (TV 1980) represented Tara with a painted backdrop. *Under the Rainbow* (Orion 1981) re-created the location, not very

In 1959 Carroll Nye, who played Scarlett's second husband in the film, and writer Norman Shavin returned to Tara for *Gone with the Wind*'s twentieth anniversary, for which these photos were taken.

well, for a farcical interlude involving the little people brought to Hollywood for *The Wizard of Oz* (MGM 1939).

The most ambitious and accurate post-Tara Tara of all was constructed, full-scale, in South Carolina for the decades-delayed sequel *Scarlett* (TV 1994). This homestead was a more-or-less exact copy of its Culver City stepsister, although it had to be re-created by that expensive project's art department without the benefit of the original set drawings, which could no longer be found, although today a partial set exists at the University of Texas in Austin. For *Scarlett*, Tara was one of the few aspects of the original production that were faithfully copied. For the six-hour miniseries, for example, Scarlett O'Hara was now played by a brown-eyed actress, Joanne Whalley.

Yet even as Tara, or at least our perceptions of Tara, splintered and reflected upon a hundred venues, the original, the "real" Tara, as noted above, continued its sad decline on 40 Acres.

When Desilu bought the studio in 1957, Desi Arnaz and Lucille Ball would have seemed to be the perfect landlords, as they were actually interested in, and a little bit intimidated by, the legacy of the property they had inherited. Lucy, in fact, had in 1938 auditioned, disastrously, for the role of Scarlett O'Hara, little realizing she would later actually *be* the mistress of Tara in real life.

In a 1958 *Atlanta Journal and Constitution* article by Norman Shavin, Desi Arnaz said:

> When we bought RKO studios last year to house our expanding Desilu Productions, we acquired a colorful page in motion picture history. Included in the sale were sets and props made immortal through scores of memorable films. Of the sets left standing, probably the most famous is the imposing plantation Tara, used in *Gone with the Wind*. When I saw Tara, on my first tour of the RKO-Pathé Studios, I was still impressed by its size and grandeur—even though it had stood unattended for 20 years. Fire-scarred, weather beaten and crumbling in spots, it nevertheless retained a great and silent dignity.
>
> There had been some talk of razing Tara, but after that visit I made up my mind to restore it to its original state. Why? Well, there was the practical reason that, so restored, it could still be of use. Actually, however, my reason was more on the sentimental side. To me Tara represented a milestone in the history of the film industry. These landmarks are few, and I think, if possible, they should be preserved to remind not only the public, but the industry itself, that great achievements should always be remembered.
>
> So we are restoring Tara as soon as the estimates and plans for its rebirth are completed and time permits us to undertake the project. It will probably cost us somewhere in the vicinity of $15,000.[11]

In spite of Desi's impassioned, ghostwritten, and yet probably sincere sentiments, the restoration of Tara never happened. Instead Desi, never one to spend unwisely on the studio, was convinced by someone, probably Julian Foster, a mysterious Georgia entrepreneur, that Tara would be better preserved if it could come "home" to the South.

So, the set was dismantled, although the brick columns, which were impractical to ship, were instead just toppled on the site and then left to decompose. The rest of the set—the ersatz walls and windows and doors—was shipped to Georgia for a proposed museum / amusement park, which Foster was never able to quite make happen. The problem with monetizing the set came from an unexpected source, the Margaret Mitchell estate. The increasingly reclusive author had died in 1949 and her brother and executor, Stephens Mitchell, stubbornly refused to license anything, including the actual movie set, which might have exploited the book's success.

In 1979 the by now physically ill, and ever more financially strapped Foster, who had been frustratingly outmaneuvered at every turn by the Mitchell estate, and specifically by Stephens Mitchell, who was himself an attorney, called Betty Talmadge, the wife of a former Georgia Senator and a prominent Atlanta socialite, and talked to her about buying the set, which was by then little more than a jigsaw puzzle of doors and windows and wood trim. The price that Foster was asking, however, was $375,000.

Talmadge did not buy Tara right away, but was haunted by the memory of seeing the dissolving flats and frames Foster had shown her mysteriously stacked in a shed. Eventually she offered the businessman $5,000, which, in desperation, he accepted.

However, Julian Foster died before she could collect the movie set from him, and Talmadge, remarkably, had no idea where the set she had purchased was being stored, as she had been blindfolded, cloak-and-dagger-like, by the paranoid and by then dying businessman, who had insisted on that precaution before agreeing to drive her out to the site for her one-and-only visit. Eventually she rented a helicopter to fly her around north Fulton County to locate the barn, which she did. Tara was hers.

Talmadge would next, in 1980, purchase the ancestral home of Margaret Mitchell's grandparents, the Fitzgeralds, and move it to a pasture on her plantation in Lovejoy, Georgia. *That* plantation, built in 1836 and complete with those all-important white columns, had, according to some sources, been Mitchell's inspiration for Twelve Oaks. But the Fitzgerald home, especially when propped on cinderblocks alongside it, looked like nothing so much as a midwestern farmhouse. "Plantation-plain," is the architectural style the Fitzgerald house most resembled, which was popular in the nineteenth-century South, but which, to the disappointment of fans of both the book and the movie, most certainly did not include those white columns. A freak tornado would destroy this homestead in 2005.

The Tara set didn't include the columns either, which were still in Los Angeles, somewhere. But what did it include? To get an answer Talmadge hired Tommy Jones to inventory her holdings and had her pile of debris evaluated. A 1992 article in the *Los Angeles Times* recounted what she learned: "Guess what [the analyst] appraised it at?" she said. "$1.2 million. I said, 'You're crazy!' He said, 'No, Mrs. Talmadge, it is the most famous movie facade in the world.' I said, 'Oh, my God!'"[12] Yet that most famous of movie sets is obviously less than complete. The front door, for example, has been separated from the rest of the collection and was restored, in 1989, by Tommy Jones, and then put on display, first at the Atlanta History Center, in 1989–90, and later at the nearby Margaret Mitchell House Museum, although it is still the physical property of the Talmadge family. Jones asserts that his analysis of the original materials during his restoration is completely consistent with the 1939 construction date and with all the historical documentation, including historic photographs that were exhaustively consulted during the process. He recalls that process and that it "took most of a year to have missing millwork replicated, to do lab analysis of what paint was left on the set, and to repair what was there, and then to put it all back together."[13]

Back in California, rumors occasionally have bubbled onto online fan sites and among collectors that those brick columns, which remember, never left the state, had been salvaged by an "anonymous Los Angeles collector." That collector, with pointed irony, turned out to be the studio's future historian, Walter O'Connor.

O'Connor was a charter member of the cult of 40 Acres. When the property was being destroyed, in 1978, he determined that he, because no one else in the world seemed to care, would be the one to document the death of a backlot.

In a letter to a friend, Nancy Knechtel, O'Connor described taking photos and making movies

> of the last remaining vestiges of the old "40 Acres," piles of debris, roads, lay of the land, etc., a last desperate attempt to record that somewhat hallowed ground for posterity before the concrete and asphalt, cancer-like, took over.
>
> I made almost daily pilgrimages to the place and went over every inch of the ground by foot. So, I came to know it all quite well.
>
> One day in the latter part of October I made another examination. My visits had grown fewer as all vestiges of the old times were completely erased and I was absolutely startled to discover several huge sections of brick columns resting above the ground where they had been unearthed from a 5–6' trench cut by a bull dozer.
>
> Needless to say, I was ecstatic, recognizing them immediately as broken remains of Tara's imposing facade: The fact that the spot where they were dug out was the *exact* site of the legendary Tara only lent more credence to their origin. . . . I quickly inquired if I might have them, never alluding to just what I

knew them to have been a part of, and was given permission by the job foreman, a Mr. Jeff Players.

I then made arrangements with a local crane operator, and very early one Saturday morning, we began their removal: this story alone is worth repeating, but I won't endeavor to do it at this writing.

Though I was never able to confirm their actual existence, I believe the remaining sections are there now, permanently entombed by those ugly modern structures which now occupy the famous site.[14]

Although O'Connor intended to rescue all of those long-toppled columns, the tractors bearing down upon the site, as well as a lack of storage space in his garage, ultimately thwarted those plans. He did keep what he could, however. These remaining bricks, along with other various other items and artifacts from *Gone with the Wind*, acquired during his days as the studio historian, as well as from other collectors, outside estate sales and auctions, are now in the possession of his heirs, who are planning to sell his extensive collection at some indeterminate date in the future through Morphy Auctions in Denver, Pennsylvania.

Tara today. (2018, photo courtesy of Peter Bonner.)

Meanwhile, south of the Mason-Dixon line, an Atlanta enthusiast, Carolyn Ashworth, now claims to own an upstairs window and two shutters, the ones that Mammy (Hattie McDaniel) called out to Scarlett from in the opening scene—although how and why these pieces were severed from the main set is, like the rest of the story, a bit of a mystery. Apparently, Ashworth had earlier tried to buy Tara from Betty Talmadge, but when Ashworth's nonprofit failed to pay for the set Talmadge repossessed it, except for that window. Tommy Jones refutes this, explaining that Talmadge originally lent Ashworth those windows to help with fund-raising for her project. The Talmadge estate, although not Betty herself, who died in 2005, occasionally has tried to discredit these pieces as not being a part of the "real" Tara.

Today, in the twenty-first century, the current curator of Tara is Peter Bonner. As Bonner puts it, "Today the Tara facade is still in Mrs. Betty Talmadge's barn in Lovejoy, and I am still working to get it into a larger building to better-display this piece of history. I have no idea when it will happen, but every day I work toward that end."[15]

When asked what else there was to say about the set, Bonner replied that "there is nothing to say other than Tara still exists. I offered to loan the front doorway of Tara to the new Academy Award Museum, and they said 'no, thank you.'"[16]

To return again to 40 Acres, and to the early 1960s, with Tara no longer on the backlot, several acres of real estate were now open for future production.

That future production would, once again, involve Jesus Christ.

Cecil B. DeMille's phenomenally successful *Samson and Delilah* (Paramount 1949) and *The Ten Commandments* (Paramount 1956), along with some other exotic blockbusters of the period, triggered an unparalleled interest in biblical themes that, with the possible exception of the Crusades era (which DeMille, of course, had also once made a movie about), was truly unprecedented in western culture. Therefore, many more biblical, and biblical-era, epics followed.

In the early 1960s, not one, but two accounts of the life of Christ would be released. One of them, *The King of Kings* (MGM 1961), a remake of DeMille's original, was produced overseas, in Spain specifically, as was increasingly the way, on an adequate but less-than-lavish budget. The other, *The Greatest Story Ever Told* (United Artists 1965), by contrast, would receive the Hollywood treatment, with a legendary director, George Stevens; an all-star cast, including Max von Sydow, Angela Lansbury, Pat Boone, Charlton Heston, Claude Rains, Sidney Poitier, Shelley Winters, Carroll Baker, and even John Wayne; and, of course, sets constructed on the very site of some of Hollywood's greatest triumphs, at 40 Acres.

At least some of the city of Jerusalem sets Stevens built on the lot were actually intended to be constructed on location—not in Jerusalem, but in Utah and Arizona—before unprecedented amounts of snow forced the beleaguered production back early to California. At the time, and sadly even more so today, Hollywood was

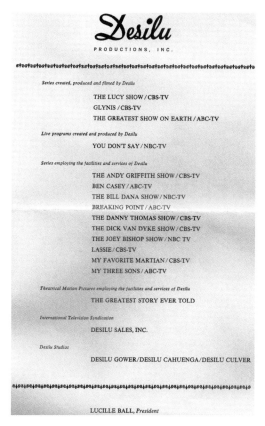

In 1963 Desilu was only able to boast one feature film shooting on its three lots—but it was a big one.

suffering from the effects of "runaway production," a term that would have been new in 1965, to describe American films shooting overseas, where costs and labor were cheaper. So, building the sets for this film in Spain, as had been done for *King of Kings*, rather than in Southern California would not have been inconceivable. Locally, both Paramount and 20th Century Fox (the original studio behind the movie) tried to attract Stevens's production dollars with their facilities, but Desilu won out, probably because of its still-lower below-the-line costs.

Once Stevens agreed to build those sets in Culver City, however, Culver City nearly toppled the deal by informing the director that he could not shoot at night because of the residential noise ordinances in effect in the suburbs around the backlot. A strongly worded 1962 memo from Stevens Productions VP Frank I. Davis to Harry Bond, Culver City's "chief administrator," sputtered out that "in order to keep the picture in the US we must have cooperation from within the industry, and also from those of us who have an interest in maintaining the industry in the country." Apparently not wishing to be blamed single-handedly for the death of Hollywood, Bond ultimately capitulated and Stevens was allowed two weeks of night shooting.

The Greatest Story Ever Told. Max von Sydow is in front of the big outdoor set on 40 Acres, although the Jerusalem street scenes that followed were shot on a soundstage. That *is* John Wayne in the foreground, right. (1965, photo courtesy of Joseph Musso.)

40 ACRES: HOLLYWOOD IS A FACADE

But the misfortunes, if not the snow, followed the production to Culver City. Acclaimed actor Joseph Schildkraut, who was playing Nicodemus, passed away (in New York) before his role was finished, and cinematographer William C. Mellor also died of a heart attack on set, during filming. The bad luck continued after the release of the movie, when unfavorable reviews and audience apathy doomed the $20 million film commercially, and single-handedly brought an end to the production of biblical-themed epics for years afterward.

It is regrettable that films set in ancient eras became perceived as box-office pariahs so quickly after *The Greatest Story Ever Told*'s release, because if the trend of screen epics had continued for even another year or so, Desilu would undoubtedly have benefited by renting the sets for Stevens's epic out to other, less prestigious and hopefully more profitable productions, and that would have ensured the survival of that set, and the sets around it on the lot, perhaps into another era, perhaps on to today. But instead, the sandstone walls and palm-lined marketplaces of ancient Jerusalem were almost immediately taken down.

What replaced these sets was a surprising, ironic, even sacrilegious flip-flop between the sacred and the profane, between the devotional and the irreverent. The land where Scarlett O'Hara once pouted and Jesus once walked was rebuilt, shockingly, into a World War II–era concentration camp—and for a prime-time sitcom yet.

Actually, as fans, and there are even today many of them, like to point out, the TV series *Hogan's Heroes* (TV 1965–71) was set in a World War II prisoner-of-war Stalag, not a concentration camp. So be it.

Stalag 13, for *Hogan's Heroes,* blended Nazi and sitcom antics. The palm trees were (usually) kept off-camera. (circa 1968)

Hogan's Heroes was, like many TV shows of its era, a bastard stepchild of a film, in this case Billy Wilder's *Stalag 17* (Paramount 1953), and of the memories of a great many WWII veterans who had returned from the war and been able to process that war only by laughing at it. So, in *this* TV-land WWII, the Germans are comic foils for the much smarter Allies, who also happened to possess, unlike their Axis foes, a keen and cynical, Billy Wilder–esque sense of humor. Unlike later, more politically correct generations, who have objected to the concept of finding humor in horror, these veterans found the antics of Colonel Robert Hogan and the

The *Hogan's Heroes* sets were still in use on October 29, 1970, when Rex McGee took these mournful studies of a familiar set near the end of its life. (Photos courtesy of Rex McGee.)

40 ACRES: HOLLYWOOD IS A FACADE

Teutonic Commandant Wilhelm Klink, and let's not forget his bumbling sergeant, Hans Schultz, to be endearing, even therapeutic—had they ever had occasion to require therapy. And so, for six years, longer than the length of the actual war it celebrated, and memorialized, and mocked, and yet never really satirized, *Hogan's Heroes* here re-created that war.

Stalag 13, which was called Camp 13 in the pilot, was one of the most impressive sets ever constructed specifically for television. Over a dozen barracks and watchtowers and administration buildings, along with a water tower and a double barbed wire fence, were all created (or adapted) on-site. Artificial snow, portrayed by salt and white paint, was dusted onto the rooftops and on the ground, to create a permanently chilly German winter. The surrounding berms concealed the palm trees and the suburbs of Culver City, which by now had completely encircled 40 Acres. Inside those suburbs' tract homes, television sets now glowed blue, transmitting images created, in some cases, just over those very berms.

A 1968 *Mission: Impossible* (TV 1966–73) episode, "Trial by Fury," used the Stalag 13 set while *Hogan's Heroes* was still on the air, this time as a banana republic prison camp. That snow therefore had to be removed from the set, temporarily, although at least the palm trees in the background no longer had to be avoided.

In 1975, four years after Hogan and his heroes had been liberated, an independent production company, Ateas Filmproduktions—an entity that never made another movie, by the way—rented what was left of the set for *Ilsa: She Wolf of the SS* (Cambist Films 1975). It used the camp to depict World War II Nazi atrocities, mostly sexual in nature, in very graphic detail. At the end of the movie, most of the set was destroyed, which probably saved the current 40 Acre landlords on the pending demolition costs. *Ilsa's* kinky adventures were successful enough to spawn several torture-porn sequels, which thankfully would be filmed elsewhere.

Back on 40 Acres, on this same site, there was still, astonishingly, one more movie waiting to be made. That movie, this time, would not be an exploitation film, but rather a big-budget affair starring two of the biggest stars in the world.

The Fortune (Columbia 1975) was a 1920s-era farce by director Mike Nichols, which spent nearly a million dollars building an apartment complex and courtyard, as well as the surrounding houses and landscaping, on top of the recently machine-gunned POW camp and populating it all with expensive period automobiles and expensive actors Warren Beatty and Jack Nicholson. Art director Richard Sylbert, who specialized in early twentieth-century settings, had fun with the 1920s milieu, although many actual Los Angeles streets just outside the gates, minus a few vintage cars and fire hydrants, even today look remarkably similar to what he built at 40 Acres.

7. Railroad Station Set (originally the Temple of Jerusalem)

This place was called the "Train Shed," the "Railway Station," the "Railroad Depot," and the "Train Yard" in studio production reports and memos—when it was called anything at all. Margaret Mitchell referred to it as the "Car Shed." Obviously, everyone on the lot already knew what and where it was without having to give it an official title. A 1940 studio map identifies it as the Railroad Station Set, however, so that is how it will be referred to here.

This set, whatever it was named, was a defining structure on the lot for more than thirty years. Yet even before it was built in 1939, this part of the lot, geographically almost in the center of the studio, was where many of the defining 40 Acres moments and movies were created.

One of the very reasons 40 Acres existed was to house the sets for Cecil B. DeMille's *The King of Kings*. The original backlot for the studio, crammed into a narrow corner of the lot where the Scene Dock is now, obviously would not be adequate to the task of re-creating the life of Christ. So, DeMille looked outside the gates.

Paul Iribe, DeMille's art director, was given the task of designing those sets outside those gates. But Iribe, a former French fashion illustrator, who had worked well enough with DeMille at Paramount, clashed with the director and was replaced by Mitchell Leisen. It was Leisen who, in August 1926, built the Second Temple of Jerusalem sets from which Jesus drives out the money-changers, and which would dominate the backlot for its first decade, just as the Railroad Station Set would for its second, and beyond.

The King of Kings, Jesus (H. B. Warner) in action. (1927)

The set, filled with columns and stairs and a big doorway (here blocked off with black drapes), allegedly measured a gargantuan 462 by 367 feet. Although the temple was supposedly an interior set, as was often still the case during the silent era, the structure was constructed outside and without a roof so as to allow for natural lighting, diffused here with translucent sheets.

According to filmed evidence, pioneering director D. W. Griffith visited the set during the crucifixion scene (reportedly shot on Christmas Eve). Decades later Griffith, by then a forgotten man in Hollywood, would return to the studio to visit Selznick's *Duel in the Sun* (SRO 1946), chat briefly with costars Lillian Gish and Lionel Barrymore, both whom he had discovered, then slip, ghostlike, away from the lights and out from the soundstage.

Another, less heralded visitor (in 1927) was author Ayn Rand, an extra at the time who somehow managed to attract DeMille's personal attention, and ultimately ended up with a job in the studio story department. Rand also managed to impress the film's costume designer, Adrian, who gave the little Russian bit-player a costume more to her liking than the rags she was originally assigned, and later would design some of her personal wardrobe. And as if all of this good fortune wasn't enough, Rand also met her future husband, Frank O'Connor, who was playing a Roman legionnaire, on that same very eventful location. DeMille, and the rest of Hollywood, seemed to be somewhat bemused by the lucky 22-year-old. Rand eventually ended up working in the costume department at RKO, before selling her first screen story, her first Broadway play, and her first novel, and then turning her considerably focused energies elsewhere.

The King of Kings set survived that film's apocalyptic finale and would remain the centerpiece of 40 Acres for the next half-dozen years, as smaller sets for smaller pictures sprouted up like dandelions around its imposing walls and columns. However, if those walls and columns were used in another film between 1927 and 1933, that film managed to hide them very well, although circumstantial, and visual, evidence suggests that *The Night of Love* (UA 1927) might just have used the staircase.

The set, however, would eventually make a triumphant return to the screen in not one, but two of the most famous films ever produced in Hollywood.

In 1932 producer Merian C. Cooper found that he had a problem. Like many producers, then as now, Cooper's ambitions were bigger than his budgets. And so he took a fateful walk across 40 Acres and found a set that could portray his great wall for his new film—a film which would eventually be named *King Kong*.

Kong's production, in the darkest days of the Great Depression, involved a more-or-less constant battle by Cooper to get his film financed by embattled RKO executives. Only the intervention of David O. Selznick, because of his faith in Cooper more than in the project itself, kept the film alive during its many months of innovative and expensive special-effects work. In a memo to one of his many bosses, B. B. Kahane, dated August 15, 1932, Cooper noted how "we have taken

The King of Kings sets in 1929, much as they would have looked when producer Merian C. Cooper happened upon them at 40 Acres in 1932.

the 'King of Kings' set, said to have cost DeMille over $1,000,000, and for $14,000 have remodeled it for use in 'King Kong.'" Actually, the DeMille temple, along with *all* other sets for that film, cost "only" $216,741.48, still a considerable sum for the period, but like most of the heroes of our story, Cooper and Selznick, and eventually even the less romantically inclined RKO board of directors, bought into the myth.

For *Kong*, the Jerusalem Temple would be converted into an exterior native village, courtesy of some grass huts and faux native tikis, scavenged from the recent RKO flop *Bird of Paradise* (1932). The flagstone temple floors, likewise, would be covered with sand and moss. The big doorway, previously leading into those ebony tapestries, would be expanded by art director Alfred Herman into an opening 20 feet wide and 60 feet tall, and then fitted with an imposing set of double doors which closed with a bolt the size of a tree trunk. A gong, like something out of a later J. Arthur Rank movie, was then mounted on the top of the doorway. Outside of that doorway, a full-size altar was constructed for the sacrifice scenes in which Fay Wray, indelibly cast as the fable's fairy-tale beauty, becomes the bride of the beast. When the scenes with the natives waving their guttering torches was shot, one executive, who had approved the purchase of the heavily insured Pathé lot, supposedly muttered, "I'll give somebody $10,000 to burn the place down."[17]

Herman, or Cooper, or someone, rather ingeniously took advantage of the unmistakable, and unhideable, "Romanesque" aspects of the set by combining the ancient with the pagan, to startling effect. In the final *Kong* screenplay (by Cooper, James Creelman, and codirector Ernest B. Schoedsack's wife, Ruth Rose), Captain Englehorn (Frank Reicher) remarks "Colossal! It might almost be Egyptian."

40 ACRES: HOLLYWOOD IS A FACADE   157

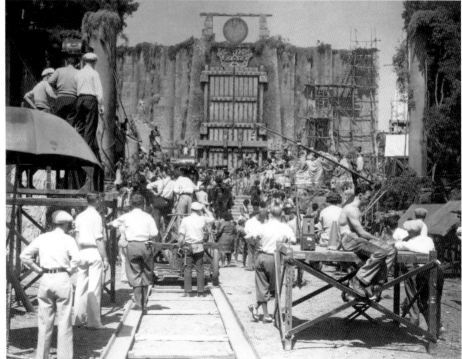

The King of Kings temple as repurposed for a different king. Note the Roman columns still standing from 1927 in both photos. (1933)

Two million feet of two-by-fours went into the making of these mighty gates, first used as part of the temple in *The King of Kings*, later battered open by King Kong, and still impressive in comparison with the six-foot man standing between them. The tablets with hieroglyphics were added for *She*.

The *Kong* wall as redressed for *She* and as described in a contemporary (1936) magazine article.

Later in the film, when Kong tries to reclaim his beautiful sacrifice, he pushes the gate open, a stunt actually accomplished by four concealed tractors on top of which Kong was added later. Schoedsack recalled that he had a "hell of a time" getting the native extras to react to nothing when the doors were pulled apart.

Post-Jesus, and post-*Kong*, the set almost immediately appeared in *The Return of Chandu* (Principal Distributing 1934), an independent serial produced by Sol Lesser, who would later become very well-acquainted with 40 Acres. *Chandu*, which would eventually, and confusingly, be adapted into two feature films, starred Bela Lugosi, playing a rare, sort of, romantic lead. The set played the "Temple of Ubasti."

The following year the Temple of Ubasti appeared in Cooper and Schoedsack's *She* (RKO 1935), which was a less successful, follow-up to *Kong*. This time the gates represented the entrance to that film's "City of Kor."

In 1936 the City of Kor made a less prominent, but more prestigious, appearance in Selznick's *The Garden of Allah*. By this time, alterations, mutations, and adaptations to the walls of the aging edifice had made the structure less Roman, less Egyptian, less Polynesian, less Arabic, and more generically exotic. This should have prolonged the facade's life for a few more years, because the less specific to a defined era, or location, a set is, the more versatile it ultimately becomes.

But this didn't happen. In 1938, Selznick's construction foreman, Harold Fenton, and Lyle Wheeler, his art director, convinced him that there was no room at 40 Acres for the acres of construction that his new film, an adaption of one the best-selling novels of all time, *Gone with the Wind*, demanded, and that many of the old sets would have to be cleared away. His crew came up with the idea of dressing some of these old sets, then burning them down for the Technicolor cameras as a set piece for his new opus, and then building the new sets on top of the ashes. The largest section of the lot to be burned was what everyone now referred to as the *Kong* wall.

As noted, most of the columns and classically old-world architecture that had long-characterized the set had been stripped away by this point. In 1938 all that

The production slate for the first Technicolor film ever exposed for *Gone with the Wind*. Part of the *Kong* wall can be seen, for the last time, in the background. (1939)

remained of the original set was a single facade that ran west and then angled in a southerly direction into a second, less lengthy yet much taller wall, where the gap from which Kong had rampaged was still visible.

The problem would be to transform this jagged wall into something that could convincingly play nineteenth-century Georgia.

The famous sequence to utilize this set would then and ever after be referred to as the "burning of Atlanta." Yet the scene vividly depicted on-screen was set in September of 1864, and General Sherman would not torch that city until November, which in both Margaret Mitchell's book and in the movie actually happened off-screen. What was actually photographed by the cameras was instead supposed to be nothing more dramatic than the burning of some munitions sheds. "Our gallant lads must of set fire to the warehouses near the depot," Rhett Butler explains in the film, unintentionally alluding to the last and next set scheduled to be built on the site.

Indeed, even if history and Margaret Mitchell had burned the city around Scarlett O'Hara, it would have been hard for Wheeler to convincingly turn DeMille's temple into anything resembling the Antebellum South. So, the first time the audience sees these unusually imposing "warehouses," they are already completely consumed by flames. Thus, anything other than the outline of the buildings themselves would have already and most conveniently been obliterated. Interestingly, the blueprints Wheeler and his staff used for the conversion were, not so cryptically, labeled "Kong set as is – 40."

Wheeler, thus freed from having to dress Skull Island for (the then uncast) Scarlett O'Hara, instead concentrated on rigging the aging facades with gas pipes so that the flames could be controlled and regulated for the cameras. Describing the scene, historical consultant Wilbur Kurtz, an Atlanta artist and academic, obviously astonished to find himself in Hollywood at all, enthusiastically wrote in his diary that "the old *King Kong* set is piped for oil sprays to make the burning a veritable inferno. A lot of wooden debris is piled behind to add its bit. A huge warehouse wall is to fall during the fire. I went down under the set to see the contraption which will pull it over. The various steel cables join near a central pulley and far out of sight line, a tractor will furnish the 'yank' that will heave over the wall."[18]

Fire and lighting tests for the scene were conducted on December 9, 1939, but the big event, the first film to be exposed for *Gone with the Wind*, was scheduled for the following night after Selznick, production designer William Cameron Menzies, production manager Ray Klune, and director George Cukor had hosted a catered dinner for the cast and crew and a platoon of local firemen on the nearby Arab Village set. Selznick and Cukor were MIA for the dinner, however, and would not show up until just before the blaze.

Only seven Technicolor cameras existed at the time, and all of them were requisitioned for the shoot. Actually, the last of them had to be rushed over on the evening of the fire from the various studios they had been working on that day. When everything was ready, there were several hundred anxious people on the set. Yet Selznick asked Klune to wait for his brother, Myron, who finally showed up, slightly tipsy, with two guests.

What happened next is, of course, the stuff of Hollywood legend. Which means that whether it happened at all has been contested ever since that evening transpired. Everyone agrees that when Menzies called action (the fire, in spite of the attention it attracted, was a second-unit sequence—meaning that it did not involve the actors—so Cukor was not wielding the megaphone that night), the flames ignited the aged set. People all over Los Angeles, seeing the blaze beneath the Baldwin Hills, panicked and reported the fire to local authorities, jamming the phone lines for hours. Many of the onlookers assumed that it must be MGM on fire, which amused Selznick, who after seeing the inferno, later that same night admitted that the sequence was "one of the greatest things I've ever seen."[19]

Something else Selznick saw that same night was British actress Vivien Leigh, who Myron pushed in front of his brother and slurred, "Here's your Scarlett O'Hara. And she's not even my client." Selznick, not an introspective man by any means, took one look at the actress, whose catlike eyes looked very green indeed that evening, and later admitted that "when my brother introduced her to me, the dying

The *Kong* wall is in the background, pretending to be Atlanta. Two stunt players are in the foreground (Yakima Canutt and Dorothy Fargo), pretending to be Rhett Butler and Scarlett O'Hara. (1939).

Occasionally, infrequently, a photograph will be taken as history happens right in front of the camera lens. Here, on December 10, 1938, David O. Selznick meets his Scarlett O'Hara for the first time. Selznick and Vivien Leigh are center (Selznick has a cigar in his mouth). A resigned Laurence Olivier is in front of them both, hands in pockets.

flames were lighting up her face . . . I took one look and knew she was right . . . I'll never recover from that first look."[20]

The cost of the whole sequence, which Kurtz romantically referred to a "Spectacle for the Gods,"[21] once the flames had been extinguished, once Scarlett O'Hara had been cast, and once the legends had themselves ignited, was later estimated as just $24,715. Today, the elaborate dinner served to the crew that night so many evenings ago on the Arab Village set, assuming it was still there, would cost more than that.

Once the ashes of the old sets had been raked away, construction could begin on the Atlanta sets, the centerpiece of which was to a reconstruction of that city's railroad depot. The original railway station, in Atlanta, had been created in 1853 by civil engineer E. A. Vincent and had been destroyed by Sherman's troops in 1864. The original was an enormous edifice, 300 feet wide by 100 feet long, and the interiors inside those walls ranged over 30,000 square feet. The backlot copy reproduced all of this faithfully, if on a smaller scale. The existing set drawings identify the movie version as being 149 feet, 4 inches long and 45 feet wide (or half size, 50% scale). Like the original, it had three rounded doorways on each end for rail traffic. On the roof, the long gable running across the entire length of the structure actually was constructed, but the curved roofline of the original was only simulated by a matte painting. Also, unlike the Dixie version, the movie set was never completed on the back or on one side—where the camera would not venture.

The *actual* Atlanta railroad station, as photographed by E. H. T. Anthony and Company Photographers in 1864, and the 40 Acres version. (1970, photo courtesy of Rex McGee.)

The primary train expected to puff its way through this set was a purpose-built prop. Selznick had been instructed by Kurtz to try to borrow an actual Civil War–era locomotive, *The General*, but the producer ultimately balked at the $4,200 it would have cost to get the train west to California (*The General* would instead ultimately journey north that year, as a star attraction at the 1939 World's Fair). Selznick was probably unaware when Kurtz made the request, but Kurtz's late father-in-law was Captain William Allen Fuller, who had recaptured the *General* from Union raiders in 1862.

On Monday, May 22, 1939, the legendary scenes of Scarlett wandering through the train shed and out into the railroad yard, surrounded by thousands of squirming, bleeding, wounded Confederate soldiers, were captured using an oversize construction crane (no film crane in the world would have been tall enough to accommodate the 90-foot elevation required). Unfortunately, it was discovered that this crane's hacking motor caused the camera on-board to vibrate. And so, the ever-resourceful Ray Klume eliminated that motor entirely by having the entire contraption, which weighed in at ten tons, instead pushed down a concrete ramp 150 feet long while, again manually, jacking the jib into the sky and over the extras. "It worked out very smoothly," Klune said modestly. Decades later, observant visitors to 40 Acres could still find traces of this cement trough amidst the weeds.

Those Confederate extras, and who really knows how many there were that day, were supplemented with another eight hundred articulated mannequins to fill out the corners of the scene. In an interview with historian Ronald Haver, Klune revealed:

> The Screen Extras Guild made a big fuss about it and tried to get us to pay for the eight hundred dummies, and I said "No, not one cent," and they raised all sorts

For the depot scene in *Gone with the Wind*, shot on May 22, 1939, the camera needed to rise more than 90 feet off the ground. This 65-ton construction crane ultimately was enlisted for the shot.

In 1939, Scarlett O'Hara crosses the Atlanta rail yard with Atlanta and hundreds of extras in the background. In 1972, Marc Wanamaker belatedly stands in for one of them.

of hell, but we held firm on it and said, if you can supply us with real people, we'll pay, otherwise we'll use the dummies. Well, they were only able to come up with a little over eight hundred, and that killed the case completely . . . We never heard another word about it.[22]

At the premiere of *Gone with the Wind* in Atlanta, Margaret Mitchell's husband was reported to have said, "If we had that many soldiers we would have won the war!"

Incidentally, some of those eight hundred dummies presumably had nearly as long a life as the set itself. A quarter century after *Gone with the Wind*, more than one cast member of *The Andy Griffith Show* recalled seeing mannequins dressed in Confederate gray stacked in a storage shed somewhere in the darkest corners at 40 Acres.

Post–Civil War, the Railroad Station Set's career was, like Tara's, less than stellar. Although *Little Men* (RKO 1940) shared that film's Civil War setting, and successfully moved the Railroad Station Set north all the way across the Mason-Dixon line to Massachusetts, that was a singular success. Overall, the set was too big to play a small-town rail depot, and it was too antebellum to play a big-city rail hub. Its physical location, however—just west of the consistently working City Streets sets—made it a popular background, if only a background, for those sets. Whether the film being shot on those streets was a comedy, a film noir, a musical, or even a Western, seeing the roof of a big industrial building in the background made audiences assume the building was a warehouse or a rail depot—which gave those sets in the foreground the lived feeling of being part of a larger world rather than something built the day before on a backlot.

The set did end up being used for a few more roles in its long second act. For example, the feature film *The Iron Major* (RKO 1947) used it, logically enough, as an 1897 railroad station. *The Miracle of the Bells* (RKO 1948) redressed it as the "Watson Coal Company," and *Where Danger Lives* (RKO 1950) redressed it yet again as a sleazy Mexican border checkpoint.

Yet the TV series *The Untouchables* (1959–63) was the production during this era that seemed the most inclined to consistently utilize the facade. Perhaps this was because much of the Prohibition-era action for that show took place around bootlegging warehouses—although in various episodes the eastern (dressed) wall also showed up as prison, a produce mart, and even as a railroad depot.

The Andy Griffith Show (TV 1960–68) used the set less often, although it stood on the edge of the very familiar Mayberry street set and can often be seen in the background if one wants to look for it. In "The Darling Fortune" (1966), for example, when the hillbilly Darling family enters Mayberry's familiar main street, the set is very visible and also, by now, very familiar. Another episode from 1966, "The

The Railroad Station Set's last act. (1970, photos courtesy of Rex McGee.)

Legend of Barney Fife," staged a comic chase scene around, inside, and in front of the building.

That building was looking pretty bad by this time. Furniture and parts of other sets were now being stored inside under the partial roof, which never offered more than token protection against the elements. The elaborate architectural patterns—the scrollwork and lattices and curlicues that had always distinguished the set, and which in a way dated back to 1853—were still largely in place, yet the "brick" around these accents was peeling away and piling up in broken heaps of plaster under the pockmarked walls like leaves dropping off of a dying tree. On the western side, the set had always been a skeleton, yet now the whole structure resembled a rotting cadaver.

Perhaps the only project to ever cast the set *as* a set was the TV series *Batman*, for the 1966 episode "Holy Rat Race" in which the Batmobile, in a chase through the "Bioscope Studios," passes the forlorn remains, and leaves them behind in the dust.

The Railroad Station Set burned to the ground in what could almost be described as a mercy killing, on the occasion of a December 1971 lot fire. The *Los Angeles Times* piece on the fire misidentified the structure, surely with no intended irony, as a "warehouse."

In the 1960s, just south of the Railroad Station Set, there was a bridge/trestle that spanned two grassy berms and was featured in television series of the time, like *Hogan's Heroes*.

8. Country Home

This surprisingly versatile set had a surprisingly long and complex life. The facade, like the alternate Farm and Barn set to the west, apparently began its life as a New Mexico dude ranch, in the funny but short-lived Desilu sitcom *Guestward Ho!* (TV 1960–61). *Guestward Ho!*, despite its rural setting, was based on a book by the colorful Patrick Dennis (*Auntie Mame*) and had a plot that prefigured the later, and more popular, *Green Acres* (TV 1965–71).

At least one episode of *The Andy Griffith Show*, regular appearances in *Kentucky Jones* (TV 1964–65), with Dennis Weaver, and an episode of *Batman* (TV 1966–68) followed. The set was still standing in 1968 when it was converted into the Sam Jones (Ken Berry) house for *Mayberry R.F.D.* (TV 1968–71). Sam's house had earlier appeared on *The Andy Griffith Show*, but had been played by a different set. The recasting of the farm involved giving this set a slight makeover, namely a new, second barn—painted red, of course—more vegetation, and smaller columns on the porch, to evoke a Southern, rather than Southwestern, ambiance.

In its last season, *Mayberry R.F.D.* moved to Warner Bros. The credit sequence from the 40 Acres version was retained, but the exterior house was now

The Country Home set as it appeared in 1970 (top) and the much later, similar but by no means identical Warner Bros. version circa 2010. (Top photo courtesy of Rex McGee; bottom photo courtesy of Mischa Hof.)

an entirely different facade. This version of Sam's house was created using a set that had already stood for several years on the inexplicably named "Doonevan Flats" backlot at the WB lot in Burbank. This "new" house, remarkably, was converted into an approximation of the old Jones farm with surprisingly few modifications needed. For many people, even in the early '70s, our perception of what a rural farmhouse looks like had already long been shaped by on-screen impressions created on backlots, rather than by life. And so one studio farmhouse tended to look uncannily like another.

As if to prove the point, this very same Warner Bros. set, shortly after *Mayberry R.F.D.*'s 1971 cancellation, was revamped yet again for *The Waltons* (TV 1971–81), when series creator Earl Hamner Jr., who based the show on his early life, noticed its resemblance to his actual childhood home in Schuyler, Virginia. This apparently last version of this set, through rebuilds, relocations, and fires, has continued to be used ever since. It has most recently appeared as the Dragonfly Inn in *Gilmore Girls* (TV 2000–7) and in an episode of *The Mentalist* (TV 2008–15).

Back at 40 Acres, another collection of barns and sheds, without a specific accompanying farmhouse this time, once stood west of this area near the northwest corner of the lot. These buildings, identified cryptically on some studio maps as the "old set," were much used in the 1960s for episodes of *Mayberry R.F.D.*, *The Andy Griffith* Show, *Hogan's Heroes*, and *Bonanza*.

9. Location Roads

"Location Roads" is a blanket term, never actually used at 40 Acres, for any of the network of trails and boulevards that honeycombed the property during this era, especially on the western end of the lot. These roads are noted here, however, because like the constructed sets they paralleled and circled around and through, they were, generally speaking, designed and used as locations, and so like those other sets, their participation in 40 Acres history needs to be noted.

These roads doubled as a way of transporting people and equipment across the lot, of course, as well as providing locations. According to Joseph Musso, who explored all of them, they were paved or partially paved, although because they were usually intended to be dusty country roads, the asphalt was often covered with dirt anyway.

To hide these location roads away from other often-adjacent locations, an old studio trick was employed. Earthen berms were heaped up along or on one or both sides of the trail, and assorted landscaping elements—grass, or trees, or rock—were employed to block out or divert attention away from the sets, or the actual city suburbs, that lay on the other side.

Almost every program, designed for big screen or small, that set up their lights and cameras on 40 Acres ended up utilizing these roads and these berms. This was

An expansive mid-1960s view of 40 Acres—and its many Location Roads.

particularly true during the Desilu era, when many of the berms were created and when it seemed like every week the Batmobile, or the Mayberry police cruiser, or *The Green Hornet*'s (TV 1966–67) Black Beauty, or Opie Taylor's bicycle could be seen stirring up the dust in front of what often looked like the same berm that Wheeler and Woolsey had raced a Roman chariot past decades earlier in *Oh! Oh! Cleopatra* (RKO 1931). Maybe it was.

The roads were intended to be generic. That was how they could be reused so effectively, for so long, although a trek down these roads during this era would have revealed, among other things, a western street adjacent to an Arab village, Gomer Pyle's Fort Henderson, Wally's Service station (where Gomer worked before his stint in the marines), the McCoy and Jones family farmhouses, the Tarzan family treehouse, Germany's notorious Stalag 13, and, of course, some of Mayberry, North Carolina. How many of us today would have just loved to have taken that trek, just once?

A hike across the rutted roads of 40 Acres in 1975, at the very end of its life.

10. City Streets

This long, approximately six-block section of real estate is sometimes referred to internally as "Concord Street." It has long been believed that this is a reference to *Little Women* (RKO 1933), but that turns out to be a myth ripe for debunking, as the surviving production records for that movie only credit one day of filming at Pathé. Most of the exteriors for that picture were actually, apparently, shot instead at the Warner Bros. Ranch.

The City Streets set apparently then was largely made up of many different sets, representing many diverse eras and varying geographic locations. These assorted, often unrelated, "neighborhoods" gradually, through invention, accidents, and modifications, eventually fused into a more or less single "city" that could reliably be used interchangeably.

That said, during the late 1950s / early 1960s period discussed here, the urban productions—those desiring a New York or Chicago look, for example—often tended to favor those sets on the western end of the district, while the more rural-based projects seemed to more often point their cameras toward those facades farther east.

Aerial view looking west, highlighting the 40 Acres City Streets district, center. (circa 1965)

Broadly speaking, the City Streets set extended from below the Railroad Station all the way to a big church facade, which was itself often visible in projects shot anywhere on the set when the camera happened to tilt east. The church, one of the largest structures on the lot and which included interiors, dated to *The Miracle of the Bells* (RKO 1948).

South of the church was a parsonage front, often used as a house, adjacent to a small section of other houses on a residential street section called, aptly, Residential Street. To the east of this section was the so-called Reform School Set and to the southwest was the Barracks Set,

The same district from the same era, mid-1960s, only closer and now looking east.

both of which will be toured separately but

40 ACRES: HOLLYWOOD IS A FACADE 173

which were often used interchangeably, or as background architecture, as needed on these familiar city streets.

This wonderful, but unintentional, verisimilitude gave the sets at 40 Acres a feeling of reality, of lives going on behind the subject being shot, which was rare on backlot sets then, more so now, and which helped make generations believe, or wish, that fictional places like Mayberry or Gotham City or Camp Henderson really existed.

The City Streets set's earliest incarnation was on the western "city" side of the district, making this the only small town that started out as a big city first. These early exteriors probably date to the first months of 40 Acres' existence as a studio, for DeMille's *Chicago* (Pathé 1927). According to DeMille's surviving records, the set—or rather all of the sets constructed for that, expensive for its time, $236,417 production—cost $29,515. And the area where these sets stood was long-after tellingly named Chicago Street, not to designate a specific geographic locale, but rather to identify the set's origins.

Chicago (as a movie, and not specifically as a set) is a property that has seen numerous incarnations over the decades—on Broadway, on film, and back again—the most recent cinematic version being a 2002 Oscar-winning musical. DeMille only produced this 1927 version, although he may have given the director credit to

The 1930 version of the City Streets, center left, was simpler and more urban than its later incarnations.

Frank Urson so as to avoid unsavory comparisons with his much-more-devotional *The King of Kings*, which had played in packed theaters earlier the same year.

DeMille probably used the set again during the remainder of his tenure; in fact, it would have been remarkable had he not. But the records of his productions, as well as of the productions themselves, are largely unavailable or long-destroyed, as is sadly common regarding *anything* produced during the silent era. It is frankly hard to fathom this today—as that era was, after all, considerably less than two lifetimes separated from our own. With the dawning of sound pictures, and of the additional manpower and paperwork required to produce those pictures, it becomes easier to track the movements across this set and those of its 40 Acres sisters, so the RKO and Selznick and (to a lesser degree) Desilu eras are easier to document.

One of the most beloved of all sitcoms, *The Andy Griffith Show* (TV 1960–68), was primarily shot at what is today known as Red Studios, but which at the time was the third component of Desilu. But that little studio, located on the southern tip of Hollywood, had, and has, no backlot. And so Mayberry, the small-town setting of the series, had to be visualized elsewhere.

Elsewhere, of course, needed to be 40 Acres.

Most backlots, historically, have always contained a small town, or Southern town, or Midwestern town setting. In fact, the larger studios still do contain such a set. The Warner Bros. version, which would eventually portray Mayberry, is, in fact, according to their operations department, annually their most rented backlot setting.

In the 1950s such a setting would have been very familiar to a majority of Americans, who would have grown up in a place quite similar to that setting. Many of these Americans would have, by this time, already left behind their own Mayberrys, seen the world (courtesy of World War II), and then fled to the cities and to the suburbs. Yet their memories of this abandoned Eden would—precisely because it had been abandoned—still burn in their hearts.

For a certain generation, the generation before the Baby Boomers and before rock 'n' roll, this image of a small town where life was simpler, yet somehow richer, would prove to be almost unbearably appealing. A look at the works of such diverse artists as Walt Disney, Thornton Wilder, Thomas Wolfe, William Saroyan, Sterling North, Rod Serling, Ray Bradbury, Norman Rockwell, Frank Capra, Booth Tarkington, Meredith Willson, and Earl Hamner Jr. reveal this longing for a place that each of them had, presumably, chosen to leave behind and then regretted it forever.

Remarkably, this sepia image of a long-gone, ever-missed, and yet simpler time proved to be so potent that the generation that created and then recycled, re-created, regurgitated, and rebooted it was able to effortlessly glad-hand this nostalgia onto the consciousness of their children, and even onto their children's children. Even today, guests visiting "Main Street USA" at Disneyland, and its assorted sequels around the world, probably cannot entirely resist feeling an errant longing

Andy Griffith and Don Knotts in *The Andy Griffith Show* episode "The Loaded Goat," 1963, and the Mayberry courthouse set as seen in 1970. (Bottom photo courtesy of Rex McGee.)

for a life even their grandparents would have recognized only from watching *The Andy Griffith Show*.

It is perhaps surprising how the set epitomized small-town America so well for so long, as it was built, largely, to portray those urban non-Edens, Chicago and Atlanta. Yet unlike many other outwardly similar sets for many other TV shows and movies—which had often been thrown up the morning the production was scheduled to be shot, and where the paint was still wet and the bricks were all square and unblemished—"Mayberry" had a lived-in feeling about it, due to Desilu's cost-cutting and to the countless modifications and touch-ups and teardowns it had suffered over its decades of loyal service. It was very much a feeling, unintentional though it was, that a real visit to a real community, like Mayberry, might have imparted to a traveler.

This Mayberry, over its life on television, first on *Griffith* and later through the follow-up, *Mayberry R.F.D.* (TV 1968–71), created one of the most fully realized of all television communities. Mayberry is so real to so many people that there have been maps published showing where every incident in every episode took place, regardless of the countless internal inconsistencies that the show itself, with its rushed production schedule and revolving-door writing staff, was never able, or interested enough, to reconcile.

For example, there is a house seen in many episodes with a leaning tree (and later, after that tree leaned too far, a tree stump), which seemingly was used in every situation where a house, other than Andy's, needed to be seen. The 1960s audiences, which were parceled out episodes one at a time, over many years, would not have noticed these contradictions, but syndication, and later home video, and later still on-demand services, would make these errors more apparent. Yet the appeal of Mayberry, and the unhurried, trusting natural order it offered, was such that none of this really mattered. In fact, Mayberry chatrooms and websites and message boards exist online that find often-ingenious ways to reconcile these errors.

Even Andy Griffith himself, as far back as the run of the show, seemed occasionally to buy into the idea that Mayberry, North Carolina, in spite of all evidence to the contrary, on-screen and off, was real. He had based the place on his own hometown of Mount Airy, North Carolina, but Mayberry must have occasionally seemed to be the inspiration, rather than the result. In 1967, Griffith mused, "Funny thing about the Mayberry we know at Forty Acres is that even though all the buildings are false fronts, when you're working there, you get the feeling of being in a small town. You forget that on the other side of the fence is one of the biggest cities in the world."[23]

In fact, Mount Airy, still a much smaller town than Los Angeles, now mirrors and ricochets its tenuous connection to Mayberry at every possible opportunity. They have a Floyd's Barbershop, a Mayberry Jail, a Wally's Service Station, and an Andy (Griffith or Taylor?) home listed as "attractions" on their website—although these real-life alternatives are presumably copies of sets at 40 Acres, which, as it's sometimes hard to remember, were never real anyway. They also quote Griffith as saying that "people started saying that Mayberry was based on Mount Airy. It sure sounds like it, doesn't it?"[24]

Interestingly, note that Griffith's statement does not go so far as to say that Mayberry *is* Mount Airy, or vice versa. Even if it sure sounds like it, doesn't it? The *real* Mayberry, as Andy Griffith knew, even if millions of fans never connected the dots, was, of course, in Culver City.

The appeal of *The Andy Griffith Show*, and of Mayberry, was such that in 1986, years after the original show had gone to syndication heaven, where the windfall profits were, sixteen original cast members regrouped for a TV reunion movie, *Return to Mayberry*. The title was unintentionally ironic, because by this time there was no Mayberry to return to, as 40 Acres had been bulldozed.

In fact, *Mayberry R.F.D.* had forsaken the set in 1970 for the previously mentioned Midwest Street backlot at Warner Bros.—although why the move was made has never been satisfactorily explained. Threadbare though it was at the time, 40 Acres would continue to be used for other projects for the next five years.

For *Return to Mayberry*, then, the obvious location would have been Warner Bros. The Midwest Street set there still survived and looked much as it had a decade and a half earlier. And after all, for *Mayberry R.F.D.*, token attempts to make the new set resemble the old had been largely successful, and it is doubtful that many viewers at the time had even recognized the difference. So, the Warner Bros. backlot, sharing at least some original DNA with the 40 Acres backlot, could conceivably have again portrayed Mayberry.

Instead the production (for probable financial reasons) looked north, to Los Olivos, California, an expensive community that was then primarily

The Mayberry, or City Streets, set. (1970, photos courtesy of Rex McGee.)

occupied by vineyards and art galleries and Michael Jackson. The main street, however, had a suitable small-town American ambiance. A re-creation of the Mayberry courthouse, built over an actual building, tried to bridge the old and the new and succeeded, at least partially. The road that originally came to an end in front of the courthouse now appeared to continue and was much wider, and inexplicably that road now sported a never-before-seen flagpole in the middle of the street.

Ultimately, *Return to Mayberry* was a huge ratings success. The critics were less taken with the production, although everyone agreed that as an exercise in nostalgia it was effective, even if, though no one mentioned it at the time, the promise of the title could no longer, technically, really be fulfilled.

40 Acres—like many places, many homes, and many hometowns—was never appreciated by those who were actually there, at least not while they were there. Andy Griffith probably understood this, if not immediately, then later, as decades burnished the memories of the place, and of the place it played. In the early 1980s he remembered 40 Acres and recalled, "It was in very poor repair then, but now I understand that it has been leveled. I haven't been out there in a long time. We used to hate to go there. Now I would love to go back one more time."[25]

The following represents a sample of some other visitors to the City Streets, to Mayberry, over its very long, and diverse, lifespan:

The Fall Guy (RKO 1930): assorted scenes.

Officer O'Brien (RKO 1930): assorted scenes.

Bad Company (RKO 1931): assorted scenes, shot as "The Gangster's Wife."

The Big Gamble (RKO 1931): assorted scenes.

June First (RKO 1931): "Ext. apartment and hotel" scenes were shot here in the first of six "Gay Girl" shorts.

The Tip-Off (RKO 1931): assorted scenes.

The Common Law (RKO 1931): "Chicago Street."

Parents Wanted (RKO 1931): "Ext. boarding house" scenes were shot here.

Beyond Victory (RKO 1931): Bill Boyd, later to be known forever after as Hopalong Cassidy, visits "Chicago Street."

Julius Sizzler (1931): "Ext. Chicago Street," used for short subject shot as "Little Sizzler."

Devotion (RKO 1931): assorted scenes.

Not So Loud (RKO 1931): short subject that used "Chicago Street."

Is My Face Red? (RKO 1932): assorted scenes.

The early Chicago-era City Streets set. (circa 1930)

The Past of Mary Holmes (RKO 1932): as New Jersey, although the location is not identified in the film.

Born to Love (RKO 1932): assorted scenes.

Young Bride (RKO 1932): shot "street" and "dance hall" scenes here under the title "Veneer."

State's Attorney (RKO 1932): assorted scenes.

The Roadhouse Murder (RKO 1932): assorted scenes.

Stealin' Home (RKO 1932): "Rufftown Hotel."

Rockabye (RKO 1932): "Cab stand" for Constance Bennett in the last film released by RKO under the RKO-Pathé trademark.

Girl Crazy (1932): "Chicago Street" is visited by Wheeler and Woolsey.

The Half-Naked Truth (RKO 1932): New York–set "Biltmore Theater" scenes.

Ann Vickers (RKO 1933): For the "ghetto street and apartment" scenes shot here, RKO rented a camera crane from United Artists.

Emergency Call (RKO 1933): "Ext. street" and "Ext. theater."

The Right to Romance (RKO 1933): "Norwich Hospital."

King Kong (RKO 1933): Assorted shots of the king in New York were taken here in September 1932.

One Man's Journey (RKO 1933): Lionel Barrymore plays a small-town doctor, and this is the small town.

Morning Glory (RKO 1933): Katharine Hepburn won her first Oscar for this film, but only one partial day of "Int. and Ext. Coffee Shop" scenes were shot here, for a scene with her and C. Aubrey Smith.

Snug in the Jug (RKO 1933): a (Bobby) Clark and (Paul) McCullough short subject shot over three days on "Chicago Street."

Bedlam of Beards (RKO 1934): another Clark and McCullough short, which utilized a rare interior-equipped set ("barber shop") on the street.

Wednesday's Child (RKO 1934): "Ext. building & fence."

The Crime Doctor (1934): "Ext. apartment house"; not to be confused with the Columbia series of the same name that was produced almost a decade later.

Let 'em Have It (United Artists 1935): alley and street sets.

The Young in Heart (United Artists 1938): columned courthouse near construction site.

The eastern side of the district, looking west. (1935)

Made for Each Other (United Artists 1939): "Ext. Office Building."

Gone with the Wind (MGM 1939): Many new sets were built or old sets resurfaced for Selznick's magnum opus. Historian Wilbur Kurtz, probably fearing the worst, was relieved to be told by Selznick that Atlanta "needn't look like a western town of the horse operas." He remarked in his memoirs that "I saw the set where Twentieth Century Fox Studios made its long shots of the Chicago fire. It was miniature, less than fifty feet in length. Our Atlanta set was something like 800 feet long (two city blocks) and some forty to sixty feet high!"[26] Some of the sets constructed or requisitioned for this movie would continue to stand on the backlot for the next thirty-seven years.

Rebecca (United Artists 1940): the "Prefecture, Day" scenes.

Little Men (RKO 1940): the Atlanta street sets from *Gone with the Wind* were here repurposed as, among other things, a telegraph office and the "Capitol Trust

The Magnificent Ambersons utilized these sets with both cars and horses. (1942)

Savings Bank," which in *GWTW* had been the "Georgia R.R. & Banking Co. Agency."

Citizen Kane (RKO 1941): assorted "street and town hall" and "Trenton Hall" scenes were shot here.

The Magnificent Ambersons (RKO 1942): "church," "ice cream parlor," "Reverend Smith's place" as well as "station" set and "office and sawmill."

So This Is Washington (RKO 1943): one of four "Lum and Abner" comedies the studio produced during this era.

I'll Be Seeing You (United Artists 1944): YMCA building.

Lady Luck (RKO 1946): "Ext. tenement."

Sister Kenny (RKO 1946): "Toowoomba street" (Australia) was the setting for this Rosalind Russell vehicle.

The Devil and Daniel Webster (RKO 1946): The southwest edge of the set was utilized as rural New Hampshire.

The church and parsonage sets, created for *The Miracle of the Bells* in 1948, as they looked in 1950 and in 1975. (Top photo courtesy of Joseph Musso and Walter O'Connor.)

The Long Night (RKO 1947): The hotel set used somewhat extensively here by Henry Fonda is featured, as were many of this set's shops and stores as well as the film's "Allegheny Square" setting.

They Live by Night (1948): "Int. and Ext. hotel lobby," "nightclub and street" scenes shot here.

Fighting Father Dunne (RKO 1948): "Street and brewery wagon."

The Miracle of the Bells (RKO 1948): Extensive "town and church" scenes were shot here at "St. Michael's," presided over by a miscast Frank Sinatra. The eventually-to-be Mayberry courthouse here played the town newspaper office.

Portrait of Jennie (SRO 1948): café exteriors.

Adventure in Baltimore (RKO 1949): "Ext. downtown street" scenes.

Bride for Sale (RKO 1949): streets and "fish market."

The Set-Up (RKO 1949): "Ext. alley and street" scenes.

Where Danger Lives (RKO 1950): "Ext. street"; the street this time is in Nogales, Mexico.

High and Dizzy (RKO 1950): the town of "Pottsville" for this Leon Errol short subject.

The Great Rupert (Eagle-Lion 1950): featured Jimmy Durante and a squirrel(!) on the street.

Macao (RKO 1952, filmed in 1950): "rear entrance, hotel."

Gambling House (RKO 1950): "Mott Street," "Bleacher Street," and "street and alley."

Walk Softly, Stranger (RKO 1950): "tobacco shop and alley."

Superman and the Mole-Men (Lippert Pictures 1951): the feature film that launched, and was ultimately edited into, the later series.

The Adventures of Superman (TV 1952–58): as Metropolis, and occasionally Clark Kent's hometown of Smallville, although for first-season episodes only.

The Adventures of Ozzie and Harriet (TV 1952–66): an occasional tenant.

The Raid (20th Century Fox 1954): The set portrayed an 1860s Saint Albans, Vermont.

Climax (TV 1954–58): A 1955 episode, "The Day They Gave Babies Away," featured the church facade.

The Night of the Hunter (United Artists 1955): The dark side of small-town life was well-illustrated by the sequence of Robert Mitchum's homicidal preacher nearly being lynched—with the large church facade judgmentally placed in the background.

Lassie (TV 1954–73): The wonder collie worked at the studio and on this set for years. From 1954 to 1957, when Tommy Rettig was the human star, the show was shot at the KTTV (Nassour) Studios on Sunset Boulevard. In 1957, when Jon Provost took over the lead, the series moved to Culver City, where it would remain through the various studio regimes, and various human costars, until 1971.

Around the World in 80 Days (United Artists 1956): used the western part of the set as the mysterious Orient.

All Mine to Give (RKO 1957): church. This was an RKO feature film version of the *Climax* TV episode "The Day They Gave Babies Away."

The Walter Winchell File (TV 1957–59): occasional, as 1920s New York.

The Real McCoys (TV 1957–63): assorted scenes.

Verboten! (RKO 1959): Released, eventually, by Columbia after RKO's collapse, this low-budget Samuel Fuller production has gained a bit of notoriety in recent years because of its very obvious use of sets soon to be seen on *The Andy Griffith Show*, here in a post-WWII German setting. Admittedly, it is a bit unsettling to see a Nazi flag flying over the Mayberry courthouse.

Man with a Camera (TV 1958–60): occasional, usually as New York, as seen by Charles Bronson in the title role.

The Texan (TV 1958–60): Again, the soon-to-be Mayberry sets, including the courthouse, appeared here, now in a western setting. These episodes were probably shot after *Verboten!*, because what appears to be paint covering up the Nazi propaganda is visible on some of the set walls.

The Untouchables (TV 1959–63): used in many episodes, almost all of which, interestingly, returned the set, particularly the urban western end, to the Chicago setting of its early days.

Window on Main Street (TV 1961–62): as the small town of "Millsberg."

Gomer Pyle, U.S.M.C. (TV 1964–69): Gomer Pyle often returned to Mayberry, although it was here usually now cast as an unidentified town in California. For example, in one episode, Mayberry was convincingly and amusingly redressed as Chinatown in San Francisco.

I Spy (TV 1965–68): used in the 1966 episode "Cops & Robbers," probably others.

Hogan's Heroes (TV 1965–71): Several episodes used the western end of the set as assorted WWII German towns, hamlets, and cities.

Ride Beyond Vengeance (Columbia 1966): again, as Coldiron, Texas.

The western "big city" end of the street as dressed for an *Untouchables* promotion. Robert Stack stands center, while the Railroad Station Set stands in the background. (1960)

The Green Hornet (TV 1966–67): occasional.

That Girl (TV 1966–67): At least one first-season episode cast the western end of the set as New York City.

Batman (TV 1966–68): seen in the episodes "True or False Face" and "Holy Rat Race" (both 1966).

Star Trek (TV 1966–69): Several episodes were shot on the street and feature strange and unintended Mayberry references, certainly triggered by the reuse of the same sets, rather than any intentional symbolism. "Miri" (1966), for example, features the familiar streets redressed and covered with dirt and debris. Visible in this episode are the Mayberry Courthouse, the alley, Floyd's Barbershop, Walker's Drugstore, the hotel, the Mayberry bank, and the grocery store. "Errand of Mercy" (1967) featured some of the same places, now portraying the planet Organia. "Return of the Archons" (1967) featured those same familiar places again, now as the planet Beta III.

Finally, the classic episode "The City on the Edge of Forever" (1967) again used those "Miri" alley sets, as well as Floyd's Barbershop—which here William Shatner and Joan Collins walk by at one point. While they are doing so, the "Floyd's" sign is plainly, and confusingly, visible on the window. In an unintentional *Griffith* crossover episode from the same period, a "21st Street Mission" sign painted for this *Trek* episode also can be seen, unintentionally, in Mayberry—although apparently no one was troubled by these alarming time-space discrepancies when both shows were still on the air.

Family Affair (TV 1966–71): Again, at least one first-season episode cast the western end of the set as New York City.

Judd for the Defense (TV 1967–69): redressed the streets to play a dizzying variety of US cities and towns. This series about a traveling Texas attorney (Carl Betz) also had two alternating interior courtroom sets, which were redressed for each episode to represent yet another city. Although in the second season these interior sets were relocated to 20th Century Fox.

Land of the Giants (TV 1968–70): The street and adjacent church played a miniature town in the 1968 episode "Ghost Town."

The Dynamite Brothers (Cinemation Industries 1974): This Kung Fu / blaxploitation flick, sometimes known as *East Meets Watts*, apparently shot a few fight scenes in a few alleys over a few nights on 40 Acres. Perhaps more significantly, this title (both titles) was probably the first example of what would soon be a sad mainstay tenant at 40 Acres—the independent exploitation film.

It's hard to judge at this late date how many of these films were shot on the lot, as records—beyond, sometimes, the films themselves—no longer exist in most instances. Leo Pepin, still the operations manager for Culver Studios during this era, mentions, it seems all too casually, a "semi-nudie now shooting" in a promotional handout for the lot in the early 1970s, which was probably not the first of its kind. No one seemed to take any particular note of the irony of this "semi-nudie" action happening in approximately the same spot as the locations for *The King of Kings*, *Gone with the Wind*, and *The Andy Griffith Show*, among hundreds of others.

Switchblade Sisters (Centaur Pictures 1975): Another outlandish exploitation picture, this one at least was shot with a degree of artistry by director Jack Hill, which, along with a certain feminist perspective, has probably earned the film its small, later-day cult following. For the film, a violent battle between two gangs was shot on the then-very-ragged City Streets set. Hill recalls that "at the time, all of these buildings had been used in maybe hundreds of movies, and here again we had the choice of shooting in an actual street or using a set. And I opted for the set, because I wanted it to have kind of an unreal feeling that you could get with a set that you can't get with a real street." Director Quentin Tarantino, a big admirer of the film, nailed the surreal, sad appeal of the scene, which, he said, "looks like it takes place in movieworld."[27]

Lepke (Warner Bros. 1975) and *The Four Deuces* (Embassy Pictures 1975): These were more upscale, but somehow less resonant. Both were created by prolific Israeli producer Menahem Golan, who appears to have had some sort of production facilities deal with Culver Studios at the time. Golan had intended to follow these two minor gangster films with a third entry, *Hit the Dutchman*, but the failure of the first two entries of his crime "trilogy" led to *Dutchman*'s postponement, to 1992.

Vigilante Force (United Artists 1976): This was probably the last film ever made on these movieworld streets. Kris Kristofferson, accompanied by a most accomplished cast (for a B-movie), rampages across the remains of the sets, now almost unwatchably tattered and blighted. In the finale, underpaid stuntmen fall out of windows and exploding buildings. By this time the set was so decrepit that it is no longer convincing as a small town. The "brick" on the set walls can be seen peeling unnaturally off of the flats, and outdoor terrain is embarrassingly visible through the "windows" of the buildings. So horrible is the condition of the facades at this time that it would have been hard to cast them even as a movie studio backlot, as

40 Acres' ultimate decomposition was rapid and alarming, as is borne out by the contrast between these familiar sets in 1970 (left) and only five years later at the time of *Vigilante Force*. (Left photo courtesy of Rex McGee.)

surely no functioning studio would have let these now-skeletal corpses decompose to such a degree.

11. Barracks and Dock Sets

The so-called Barracks Set stood at the very bottom of the City Streets section, and despite its mysterious and unexplained name, it usually played not a military barracks at all, but rather a seedy industrial area, a waterfront, a dockyard, a border town, or an auto yard. A small cement basin, known as the Dock Set, stood nearby for a time, which can be seen in *Intermezzo: A Love Story* (United Artists 1939) and *Johnny Angel* (RKO 1945), among others.

The set was dominated by a large clapboard structure that sometimes played a house, but more often was a wrong-side-of-the-tracks type of business: a warehouse, clubhouse, or cathouse, for example.

In the 1960s the Barracks Set was still standing and being utilized occasionally, but the Dock Set had been drained. Although the basin was still partially visible, most, if not all, of the nautical set dressing had been stripped away.

Additionally shot on the Barracks Set:

The Monkey's Paw (RKO 1932): An "Eastside Brewery" set is thought to have been constructed at the Barracks. If so, this is the set's earliest known appearance, although the dock seemingly appeared in some earlier, DeMille-era nautical projects.

Westward Passage (RKO 1932): No stage or locations were listed in the production records for this (very early) Laurence Olivier vehicle, which was actually an Ann Harding vehicle. 40

The somewhat mysterious Barracks Set. (1952, photos courtesy of Joseph Musso and Walter O'Connor.)

The Dock Set as utilized in *Intermezzo: A Love Story*. (1940)

Acres *was* a location, however, and very flimsy circumstantial evidence ties the film to this set.

Intermezzo: A Love Story (United Artists 1939): as "Cagnes-sur-Mer."

The Ghost Ship (RKO 1943): assorted scenes.

Morning Becomes Electra (RKO 1947): "waterfront."

Station West (RKO 1948): office, sawmill, and warehouse scenes shot near here.

They Live by Night (RKO 1948): "border town."

The Miracle of the Bells (RKO 1948): To create "Coal Town," which the set was often thereafter referred to, a railroad track, warehouses, and tenements were added to the set.

Jet Pilot (RKO, shot in 1950, released in 1957): as a dangerous Russian border village.

Gambling House (RKO 1950): "streets near dock."

Blackbeard, the Pirate (RKO 1952): The neighborhood is again heavily redressed as a "Plaza Dock."

The more-urban district beyond "Coal Town." (1955)

The Real McCoys (TV 1957–63): The still-standing "Coal Town" is used in the 1963 episode "Aunt Win Steps In."

The Untouchables (TV 1959–63): used several times.

Guestward Ho! (TV 1960–61): used at least once.

The Andy Griffith Show (TV 1960–68): Mayberry's finest apparently patroled these neighborhoods. "Coal Town" was, according to the production records, utilized in 1963's "Man in a Hurry" and the same year's "High Noon in Mayberry"; although the episodes themselves admit to no such thing. Perhaps these records were referring to the perimeter area or the adjacent road, or perhaps Coal Town was used to park trucks or equipment.

Careful, My Love (TV 1963): A failed sitcom pilot, titled "Hide and Seek," shot in "Coal Town."

Gomer Pyle, U.S.M.C. (TV 1964–69): used at least once.

Hogan's Heroes (TV 1965–71): used at least once.

Mission: Impossible (TV 1966–73): occasional.

12. Residential Street

Northeast of the City Streets set, behind and parallel to the church, for decades rolled a suburban street with several rows and partial rows of attractive one- and two-story houses that wrapped down into the eastern corner of the lot and around the back of City Streets' impressive church facade. Some of these structures were relatively complete; others were only two- or three-sided. The houses near the eastern end of the "suburb" tended to be the most complete, as they were used most often and afforded the greatest variety of angles for the cinematographer, who when pointing his end west, could make the street look much more extensive than it actually was.

The Residential Street ran east past the church, which would be on the left. The Andy Taylor house is on the far right. (1969)

The most famous house on this street was the Sherriff Taylor house from *The Andy Griffith Show*. There were altogether six structures on the northern side of the street. Andy's was fourth going east, and so the multiple, much-beloved scenes of him and his extended family on the porch could be shot there, or on a soundstage reproduction, with the camera looking north into that porch, or west into Mayberry. Historian Kipp Teague tells us that the house originated as an interior set in *The Girl Most Likely* (RKO 1958). A view looking farther east would have, at least in the early days, revealed the gutted Reform School Set still in the background.

Next door to Sherriff Taylor's house, still looking east, was the facade known internally as "Aunt Pittypat's House." One of the few structures on 40 Acres afforded a consistent name, this odd, Victorian-era home was, of course, named after the fluttery Laura Hope Crews character in *Gone with the Wind*.

The house always looked like it had been constructed without a roof, which was perhaps intended to be added later optically. Yet, in most of its on-screen appearances, the set was photographed with its blunted top clearly visible to the audience. For example, in a 1963 episode of *The Andy Griffith Show*, "The Haunted House," the structure is shown full-on, albeit covered with vines and dead trees, but with that flattened roofline, playing the "old Rimshaw place."

Incidentally, in other episodes of the same show, the same house, with the same roof, but without the vegetation, also played the mayor's residence. In reality, the building did have a partial roof, which went across the top of the structure for a foot or so from the outer walls before stopping, presumably, out of camera range.

Additionally shot on the Residential Street:

Open House (RKO 1931): short subject featuring "fraternity houses."

That's News to Me (RKO 1931): another short subject featuring the "Burke house."

She Snoops to Conquer (RKO 1931): another short featuring "exterior apartment house."

Long Lost Father (RKO 1934): "Ext. Tony's house and garden wall." John Barrymore played the title role—he played a lot of long-lost fathers during this era.

Nothing Sacred (United Artists 1937): Small-town life was not always depicted with a Norman Rockwell patina. For example, the funny bit where a feral little boy bites Fredric March on the leg was shot here, amid the familiar picket fences and tree-lined sidewalks.

Banjo (RKO 1946): assorted scenes.

Fighting Father Dunne (RKO 1948): One house on the Residential Street here played the "Selby Place."

Aunt Pittypat's house in the 1960s and in 1975.

Adventure in Baltimore (RKO 1949): "Ext. Residential Street." Robert Young, who would later shoot a couple more movies and a series on the same set, here plays the father of an aging Shirley Temple.

She Couldn't Say No (RKO 1952): Arkansas this time.

The Adventures of Superman (1952–58): used a house south of the church in the 1952 episode "The Deserted Village."

The Rocket Man (20th Century Fox 1954): assorted suburban antics.

The Real McCoys (TV 1957–63): assorted scenes.

The Untouchables (TV 1959–63): occasional.

My Three Sons (TV 1960–72): occasional, including 1961's "The Horseless Saddle," which used the same house south of the church Superman had visited a decade earlier.

Window on Main Street (TV 1961–62): Robert Young returns. Here he actually lived in Sherriff Taylor's house, relocated to someplace called Millsberg, at least for the run of this short-lived series. Also, Aunt Pittypat's house, which would later be haunted in *The Andy Griffith Show*, had a ghost problem here as well, in the 1961 episode "Haunted House."

Batman (TV 1966–68): Aunt Pittypat also would have been panic-stricken to learn that her house was now the home of the villainous Black Widow, played by Tallulah Bankhead, no less, in the 1967 episode "Caught in the Spider's Den."

Mission: Impossible (TV 1966–73): occasional.

Land of the Giants (TV 1968–70): 1968 episode "Ghost Town."

13. Reform School Set and Bombed Town (and later Alley)

One of the most distinctive structures to be found behind the gates at 40 Acres was this imposing prison, occasionally referred to as the "State Reform School" compound, built by DeMille in 1929 for *The Godless Girl*. The $722,315.17 production spent $71,330 on set construction, which this structure would have taken the majority of.

Visually, the structure looked like a combination school and prison, which, of course, was DeMille's intent. It was used fairly often in the 1930s, but the structure was more-or-less neglected thereafter, as its usage tended to be limited to period, specifically European-set, projects, even though *Godless Girl* was neither.

As a result of those years of neglect the school eventually took on the visage of a ruined European village. This look was enhanced by other facades built early on

Bombed-out European towns such as this one would have been a sadly familiar view to audiences who had survived World War II. The version depicted here, however, was a set. (1960 and 1969)

around, inside, and in back of the Reform School Set compound, for pictures like *All Quiet on the Western Front* (Universal 1930), *Beyond Victory* (RKO 1932), and *Born to Love* (RKO 1932).

During World War II, with a few notable exceptions, pictures about actual combat were infrequent because Hollywood was trying to keep up the spirits of anxious wartime audiences. But after peace came in 1945, war films marched back into vogue, and so these same sets were dressed and altered further to resemble the bombed-out European cities, which the recent veterans had liberated, and which their families had anxiously watched in newsreels during the conflict. Eventually, of course, these aging sets became the very ruins they had previously been impersonating. The imported debris became the real thing. Illusion became reality.

The Reform School Set, or whatever was left of it at the time, was ultimately demolished in the early 1960s. Some of the bombed town facades behind the set continued to be seen on the lot and on-screen for several years after that, however.

The bombed town visible, circa 1952, from outside the studio on the eastern tip of 40 Acres. (Photo courtesy of Joseph Musso and Walter O'Connor.)

By 1968 even this was gone and the land at the eastern edge of the lot was mostly now fallow. 40 Acres' days as a backlot where new sets were built were now over.

Almost.

Star! (20th Century Fox 1968) was a big-budget, Julie Andrews musical, set, like *The Fortune* (Columbia 1975), in the 1920s, and which, according to the film's publicity, utilized almost two hundred locations. One of those locations, an alley, was built at 40 Acres, approximately where the Reform School Set had stood.

For *Star*, art director Boris Levin constructed a London-set alleyway from scratch on 40 Acres. He is seen here (right) conferring with the film's director, Robert Wise. (1968)

The alley was just that—it was a paved street with a row of brick flats on either side. But period traffic had to be seen on both ends of the alley, while Andrews, playing actress Gertrude Lawrence, played cricket, while pregnant, while waiting to go onstage. Fox at the time had little room on their lot for exteriors, their backlot having been demolished in April 1961 to make room for the construction of Century City. So, 40 Acres, which at least would have been a known entity to many on the crew, was rented for the occasion.

Many on that crew must have been shocked at how ragged the lot, which George Stevens had shot *The Greatest Story Ever Told* on just three years earlier, had since become. This curious alley, which had no buildings surrounding it, would continue to stand on the site for the rest of the studio's life, and would even be used again for the less-prestigious *Switchblade Sisters*.

Other projects that shot on this site, using one or more of the above-mentioned sets, include:

The Leatherneck (Pathé 1929): assorted scenes.

Against the Rules (RKO 1931): short subject that used the *Godless Girl* sets.

Bed of Roses (RKO 1932): assorted scenes.

Ann Vickers (RKO 1933): "Copperhead Gap Prison."

The Count of Monte Cristo (United Artists 1934): assorted scenes.

Gone with the Wind (1939): The Reform School Set is seen, obscured to the point of unrecognizability behind the Butlers while they are pushing a baby carriage on Peachtree Street.

Clark Gable and Vivien Leigh take a stroll on Peachtree Street for *Gone with the Wind*. Note the already venerable Reform School set behind them, which was not nearly so visible in the film itself. (1939)

The Story of G.I. Joe (United Artists 1945): as a WWII European bombed town.

Berlin Express (RKO 1948): The bombed town was again featured in the first postwar film to be made (partially) in Germany.

Tripoli (Paramount 1950): The bombed town area is now 1805 Libya.

Jet Pilot (RKO, shot in 1950, released in 1957): again, the bombed town.

Eight Iron Men (Columbia 1952): as a WWII bombed town.

The Raid (20th Century Fox 1954): The Reform School Set plays an 1864 Union prison.

Attack (United Artists 1956): as a 1945 European bombed town.

Screaming Eagles (Allied Artists 1956): bombed WWII town.

Verboten! (RKO 1959): WWII bombed town.

Tank Commandos (American International 1959): probably the last view of the Reform School Set in a feature film.

The Andy Griffith Show (TV 1960–68): The Reform School Set can be seen, occasionally, at the end of the Residential Street in the early years of the show.

My Three Sons (TV 1960–72): The Reform School Set can also be seen here in at least one early episode.

CHAPTER 5

The Media Campus

King Kong Gone
—JIM HEIMANN–CREATED PUBLIC ART ON THE 40 ACRES SITE

AS THE 1960S noisily jackhammered around her, Lucille Ball discovered, perhaps understandably, that post-Desi, she was very much tired of being a stern studio boss. Instead she longed only to retreat into her much-loved, and much more scatterbrained, 1950s-era "Lucy" character. So, in 1967, as her second TV show *The Lucy Show* (1962–68) was winding down, it suddenly seemed like a good time to sell Desilu to Paramount—for what ultimately turned out to be a lucrative $17 million stock exchange deal.

Originally it was intended that, as a very large shareholder in the company, Lucy would continue to be involved in what would eventually become Paramount Television. But her originally congenial relationship with Paramount chief Charles Bluhdorn rapidly cooled, and so she increasingly distanced herself from both Bluhdorn and the former Desilu as well.

At the time of the sale, Desilu included not only the Culver City properties, but also the RKO lot on Gower Street and the Desilu-Cahuenga Studios a few blocks away. Paramount ultimately kept the Hollywood RKO property, which could easily be merged with their own next door, but sold the other plants off, as required by the United States District Court, which insisted the studio divest itself of the other studios. Nonetheless, Paramount's name, briefly, was installed in front of the Mansion.

It was the following year, in 1968, that the charmingly named Perfect Film and Chemical Company, whose holdings also included, among other things, publishing,

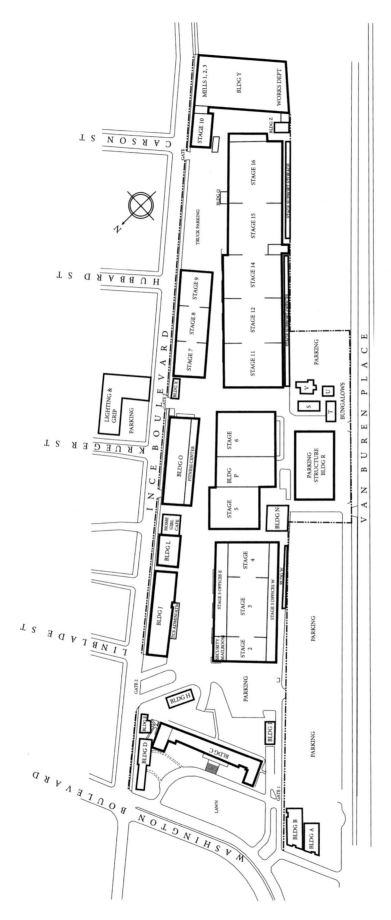

The still-vibrant Culver Studios as seen in the twenty-first century, more specifically, as seen in 2017, just before many of the pictured stages and bungalows were moved or scheduled to be repurposed. (Map by Robert Lane.)

In 1967 Paramount's name, briefly, went up in front of the Mansion.

plastics, and photography, bought the Culver City property from Paramount for an undisclosed amount. It was presumably this last interest, specifically their film-processing division, that brought them to Culver City.

Culver City and the Perfect Film and Chemical Company was not a good fit, however, and in March of 1969 the property was sold again, this time to something untrustworthily called Roberts Reality of the Bahamas LTD, a subsidiary of a Canadian company, OSF Industries LTD. The lot, purchased for a reported $9.25 million, was then renamed, inexplicably, the Beverly-Culver Studios.

Martin N. Leeds, a former Desilu executive, was the president of the new concern. He was joined in the endeavor by actors Hugh O'Brian, who was listed as vice president, and Douglas Fairbanks Jr., the deputy chairman. When pressed as to why the studio, in Culver City, bore the name of nearby Beverly Hills, O'Brian joked that "at first I was tempted to call it Mammoth Pictures because that's the name you see in movies and television when they depict a studio."[1]

Beverly-Culver Studios was apparently organized to produce original programming, including music, through a subsidiary, not surprisingly named Beverly Hills Records—which still exists—but almost immediately the board of directors decided to allocate the property as a rental lot instead. They rechristened it the Culver Studios in 1970, although it was sometimes also referred to during this era as the Culver

THE MEDIA CAMPUS 205

City Studios. It is the Culver Studios moniker, however, that the lot retains today, perhaps the only lasting legacy of that era.

About this time (1970) Eugene Hilchey, who had a history with the lot and was then running a company called Dial Effects, which created miniatures and pyrotechnics for film productions, got the idea of starting a special-effects attraction on 40 Acres, which would be open to the public and similar in some ways to the popular Universal Studios tour. Leo Pepin, who was still trying to find ways to keep the property open and viable, was agreeable, although sadly nothing ever came of the project.

In the 1970s another strange series of fires plagued both the studio proper and 40 Acres. The first was in 1971. The press reported that "two soundstages and a large warehouse on the back lot was destroyed,"[2] although the fire was, in fact, at 40 Acres and the "warehouse" was the Railroad Station Set built for *Gone with the Wind*.

In 1933 native warriors atop Stages 11, 12, and 14 waved their torches at the imminent approach of King Kong. Another fire threatened the same building in 1974.

In June 1974 another fire attacked the Van Buren Place wall of the main studio. The blaze burned threateningly right up to the wall of Stages 11, 12, and 14 (as noted earlier, there was, and is, no Stage 13). Ironically, the roof of this building had once been used as the top of the great wall in *King Kong*, from which dozens of costumed extras had furiously waved blazing torches. Fortunately, the Culver City Fire Department was able to extinguish the fire, quite literally at the soundstage door.

The next fire attacked 40 Acres in October of the same year. This one destroyed the end of the Western Street Desi Arnaz had been so excited about back in 1956. The *Los Angeles Times* rather cynically reported that the set "had not been used for at least five years and was so run down the studio indicated it had not yet estimated the value of the loss."[3]

In January 1975 yet another fire "destroyed about 100 yards of old film sets and blackened 5 acres of brush."[4] None of the media coverage of the fire mentioned which sets had been destroyed, although photographs of the fire indicated that some of the Mayberry sets from *The Andy Griffith Show* were destroyed at this time.

Hollywood in the 1960s and into the 1970s was a very different animal than it had been during the freebooting, wildcatting eras of Thomas Ince and

The results of the January 1975 lot fire included the destruction of the long-lived Taylor house, center. The ruins of the Aunt Pittypat house next door survived, however—for the moment.

Cecil B. DeMille. Fewer films were being made, and less people were seeing them. The audiences for these new films also tended to skewer older, because young people were finding even less relevance in films made by people their parents' age than they were in their parents themselves. Consequently, the seven remaining, floundering studios were frantically flailing to attract larger, younger audiences by using ultimately ineffectual tactics, such as shooting exteriors on actual locations as opposed to inside soundstages or outside on their backlots. As noted, 20th Century Fox had sold off most of its enormous backlot holdings in the early 1960s, to enormous, if short-sighted, profit. MGM, almost within sight of the Culver Studios Mansion, did the same thing in the 1970s. And the owners of that mansion could not have failed to notice the profits that resulted in those sales.

In the early 1970s, 40 Acres was still, at least outwardly, a viable business concern, but an air of slightly stale nostalgia and impending doom already enveloped the property. A 1970 piece in the *Santa Monica Evening Outlook* at the time well reflected the property's impending fate:

> The wind stirs and barely moves a dusty, faded curtain. Sunlight streams through a hole in the roof of a cavernous old building and touches a crutch here and an old canteen there. The musty smell of age permeates the air, and a blanket of dust and hundreds of cobwebs shield the building and its contents from the present. If you stand and listen long enough, you can hear the moans of wounded soldiers and the puffing of an old steam engine coming to a halt on the tracks just outside. These are the sounds of 1937 [sic]—the sounds of the railroad scene in one of the greatest movies ever filmed—*Gone with the Wind*. And you are now standing on the backlot of what is now known as the Culver City Studios.[5]

The same article went on to quote the very harried Leo Pepin, who admitted that "the property is up for sale. Today, two television shows, *Lassie* and the exteriors for *Hogan's Heroes*, are being shot on what is now known as Culver City Studios. Except for these two shows and occasional commercials, the once-busy lot is quiet."[6] Pepin admitted that he hoped to keep the lot going but that the recent trends in the industry had cast a shadow over his efforts. The article concluded by reflecting that "as you walk about the old lot and hear the history of the stages and sets, you can't help but think, 'is progress worth losing this?'"[7]

On August 22, 1974, the *Los Angeles Times* carried the first of several cryptic articles regarding the fate of 40 Acres. The first one reported that the Culver City planning commission had recommended that the parcel be rezoned from low-density residential to industrial park. This announcement is surprising because the change from studio to residential use had not been reported in the press, and this

A gallery of impressions of 40 Acres, taken in 1974 and 1975, shortly before its destruction in 1977.

was apparently the first time the issue had ever been brought up in a public forum. Even more mysteriously, in May of 1975, more than *nine months* later, it was reported that the city council had contradictorily approved the rezoning of the same parcel from "S-1 (motion picture studio) to M-1 (light manufacturing)."[8]

Later that month, on May 20, it was reported that a regional business park had been approved for the site. The article stated that the park would be developed by the "Carl M. Beck Building Company of Los Angeles for the owner, Culver City Studios Incorporated,"[9] which meant that the owners of the property, then the Beverly-Culver Studios, or whatever they were calling themselves that day, were, in fact, the culprits actively trying to amputate the backlot of their own property in anticipation of a sale, while keeping the slightly more profitable nearby stages and offices intact, at least for the moment.

In 1977 Culver City attempted to establish a movie museum in the Mansion building. City councilman Paul A. Jacobs told the press that actress Debbie Reynolds and Daniel Selznick, David's son, had "expressed an interest in the project."[10] As would happen time and again when such a project would be proposed in the future, nothing ever came of the idea.

Instead, at the end of that year, Laird International Studios purchased the lot from the ambiguous Beverly-Culver Studios. Laird had, in fact, been created by a parent company, Kings Point Corp., to manage the studio as a rental facility. The man behind the deal, Joseph R. Laird, is one of the unsung, and misunderstood, figures in the story of the studio.

The prevailing opinion on Laird, if anyone bothers to voice an opinion on the man at all, is that he was yet another in a revolving door selection of dishonest businessmen out to milk the lot in its twilight for whatever it was worth, or that he was, at the very most, an honest businessman who had the exact same goals in mind for the place. Actually, Laird was neither. He was instead a man who had, against his better judgment, fallen in love with what was supposed to a dispassionate business venture.

Laird, in fact, became so fascinated by the romantic story of the place that he forgot to carefully study what he was actually buying. He assumed, reasonably, that 40 Acres would be included in the deal. But apparently, perhaps because Laird neglected to read tiny notices in the *Los Angeles Times*, he was surprised to discover that the property had, in fact, been presold to a developer,

This far-below-the-radar sale actually involved not one, but two parcels of land: "BELOW Ballona Creek"—meaning the "old orchard," or rather "Tarzan's Jungle"—and "ABOVE Ballona Creek," which indicated the rest of the property. It appears then that Beverly-Culver Studios had secretly made *three separate* real estate deals for the same property.

Learning of this skullduggery, "Old Man Laird," as he was affectionately called, complained bitterly to those around him about the unnecessary nature of all this

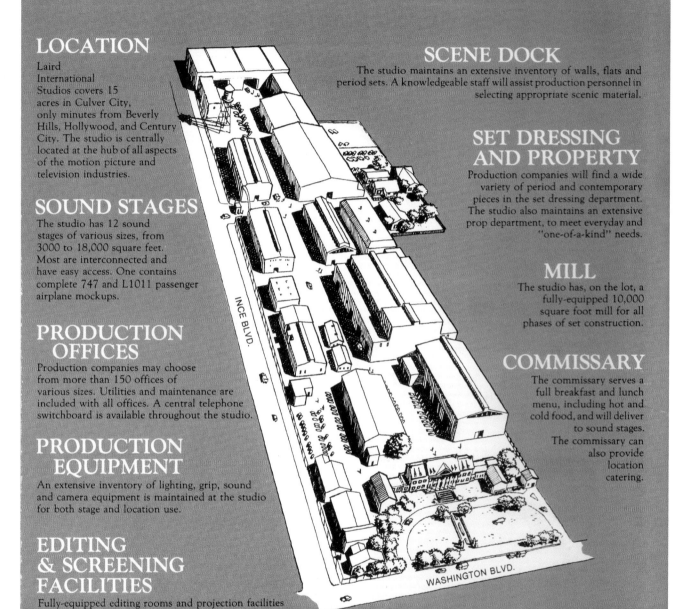

Laird International tried to make the studio an economically viable concern, and failed. (1980)

deceit—as he would have paid more than the two other buyers for the combined property, and would have kept the backlot intact and operational for the entertainment industry's continued usage.

Apparently no one listened, or cared, because 40 Acres was, it turned out, already ominously being "redeveloped." That developer, the aforementioned Carl M. Beck Building Company, almost immediately brought in heavy equipment and started bulldozing the remaining facades at 40 Acres.

Surprisingly no one, except for Old Man Laird, bothered to notice. Not a single newspaper or magazine, not the *Los Angeles Times*, not even the venerable *Hollywood Reporter* or *Daily Variety*, bothered to cover the destruction of some of the most recognizable sets in motion picture history.

Laird, hemmed in by his unrequited love for the property, soldiered on as best as he could. In fact, a new and unexpected profit stream during this era turned out to be what would become known as "music videos." Pop stars needed a place to rehearse and then shoot those videos, and they needed to do these things in secrecy and in privacy, away from the eyes of their fans and the press. The Culver Studios, because of its relative obscurity and because it offered a walled, self-contained campus, turned out to be ideal for the task. Michael Jackson shot four music videos on the lot, Madonna eight, and the Eagles, David Bowie, and Paul McCartney also used the studio regularly during this period.

Laird also rented his stages to sometimes-impoverished independent producers. He occasionally lost out on any potential profits by giving the store away to those producers, for projects he felt worthy of the still-considerable resources of his studio.

Laird's passion for the lot was so intense that he even hired a full-time studio historian, which is a position most of the majors have usually not felt particularly compelled to fill, yet he actually put someone with that title on the payroll. Walter O'Connor, the lucky custodian, was an already-eminent historian who had recently spent months, of his own accord, wandering 40 Acres in its waning days, even rescuing Tara's toppled-columns. O'Connor would ultimately become the first individual ever tasked with unraveling the studio's past, even while Laird, already floundering knee-deep in red ink, tried to chart its increasingly uncertain future. O'Connor, naturally, clashed with Jack Kincaid, the lot manager during the era, who had been tasked with the much less romantic, but perhaps more vital, job of renting the sets and the offices out, and of making sure that the commissary and the service departments were making a profit during this era.

It didn't work. Old Man Laird, sadly but not surprisingly, filed for bankruptcy protection in 1982, although it would ultimately require four years of reorganization under Chapter 11 bankruptcy laws to untangle the company's holdings so that the lot could be sold to satisfy Laird's many creditors. The last of the Kings Point assets would not be liquidated until 1992.

In 1986, after numerous legal efforts by Laird to block the sale had been finally settled, the property was purchased by Gannett, the nation's largest newspaper publisher. Gannett had recently partnered with legendary television producer Grant Tinker, who had created a string of 1970s hits beginning with the *Mary Tyler Moore Show* (TV 1970–77), but who recently had wandered into a creative dry spell.

Tinker remembers his introduction to the studio:

> I'd never been there, and one grey, rainy Saturday afternoon in October 1986 I went over to have a look. In a golf cart that had seen better days, I was piloted around puddled potholes to twelve soundstages, a number of support buildings, and some bungalows that served as office space. The last stop was the lot's signature building, an antebellum structure rather grandly called The Mansion. Like the rest of the studio, it was badly in need of some loving care. The owner had encountered financial difficulties serious enough to have the property scheduled for auction, and the entire lot was held together with spit and bailing wire. I marveled that it was a going operation, but at the same time there was something magnetic about it. The place had real restoration possibilities, all of which were later realized.[11]

The final price at auction for the then-14-acre lot was $24 million, although in his autobiography Tinker remembers being willing to go as high as $32 million

Grant Tinker Gannett, the "GTG" on Stage 2, would not last long, although this 1986-89 regime would alter the physical lot forever. (1987)

THE MEDIA CAMPUS

The construction of the new Stages 5 and 6. (1987)

to secure the property. In fact, it was reported in the press that Curtis B. Danning, bankruptcy trustee for Kings Point Corporation, had expected the bidding to go as high as $30 million. However, producers Marvin Davis and Aaron Spelling, Tinker's chief competition for the keys to the Mansion, inexplicably folded early on during the sale. The *Los Angeles Times* later reported that that bidding "was affected by the high cost of renovating the studio to industry standards and those set by Culver City officials. Estimates by movie industry analysts and others for improvements to the studio, including demolishing and refurbishing buildings, have ranged from $6 to $22 million.[12]

It is not known exactly how much Tinker and Gannett ultimately spent on these renovations (Tinker estimated it was $40 million; other reports place the number at $26 million). But the regime did add two audience-ready TV soundstages, the current numbers 5 and 6, and purchased adjacent space across the street for offices they mostly never got around to building.

Following Tinker and Gannett's acquisition of the studio and with Laird now gone, lot manager Jack Kincaid fired Walter O'Connor, the lot historian. However, when Tinker requested some historic data from Jack Kincaid regarding his proposed renovations, the pragmatic Kincaid was, not surprisingly, unable to provide that data. So when Kincaid, hat undoubtedly in hand, contacted O'Connor regarding this information, O'Connor was afforded the rare pleasure of telling his former antagonist that *that* information still belonged to Joe Laird—and so to take it up with him.

That same year, 1986, the Culver City Historical Society placed a bronze plaque on Washington Boulevard, in front of the Mansion, with the following inscription:

> On this site in 1919, the Colonial Administration Building of what was to become one of Culver City's three major motion picture studios was completed. Built by and first producing film as the Thomas H. Ince Studios, it later became known as DeMille Studios, RKO, Pathé, RKO-Pathé, Selznick, Desilu, Culver City Studios, and most recently Laird International. On the back lot, such famous scenes as the "Burning of Atlanta" for *Gone with the Wind* were filmed.

As is always the case with bronze plaques, the information contained on this one would, of course, be out of date even before its unveiling. But in live-for-the-moment Hollywood, these words, and this plaque, may have just represented something close to immortality.

Tinker was an interesting choice to manage the studio in that he would be the first executive since David O. Selznick to attempt to manage that studio as a factory for his own creative vision. That vision, however, would be focused on television rather than feature films.

The studio has always, since its chaotic founding, been a haven for mavericks and rebels and iconoclasts who have tried to make the place a creative haven, and yet instead, ironically, paradoxically, ultimately, had been trapped by the financial pressures inherent in managing that capitalist/creative haven. So like Ince, DeMille, Joseph P. Kennedy, David O. Selznick, Old Man Laird, and that endless procession of RKO executives who tried, perhaps without realizing they were trying, to create that ultimate fusion between art and business, between creativity and commerce,

An aerial view of the lot in 1986 around the time of that era's renovations.

Tinker would ultimately fail. But the studio, ultimately, would feed, vampire-like, off of all of them—and again, would survive.

The letters "GTG" (Grant Tinker Gannett) above the door of the Mansion would themselves become very quickly dated, and would be painted over in 1989 when Tinker and Gannett ended their partnership after collaborating on five consecutive financial and critical small-screen failures: *USA Today* (1988–90), *Raising Miranda* (1988), *The Van Dyke Show* (1988), *TV101* (1988–89), and *WIOU* (1990–91). Their ill-fated partnership, it should be noted, also conjured up one

216 HOLLYWOOD'S LOST BACKLOT

more critical failure, which was at least an overwhelming popular success, in a too-late-to-help first-run syndication reboot, *Baywatch* (TV 1989–99). But even the impressively toned talents of that series' stars, Pamela Anderson and David Hasselhoff, were not enough to save the company.

Tinker publicly said that he would have liked to find another partner to run the studio with, but instead his recent run of bad luck continued, and so ultimately none was forthcoming. He eventually drifted into semiretirement while Gannett, still stinging from the financial consequences of their unhappy Hollywood sojourn, searched for a buyer. That buyer would ultimately come from Japan.

And from two blocks away.

Sony, the Japanese electronics giant, had long looked west and searched for an entrée into Hollywood. In 1989 they found that special key by purchasing Columbia Studios, one of the legendary seven sisters that had presided over the American film industry since the 1930s. Since 1972 Columbia had been cohabiting with Warner Bros. in Burbank, but after Warners acquired MGM's old Culver City lot in 1989, a deal was stitched together to sell that property to Columbia in order for Warners to reclaim its original studio.

Sony, at the same time, was spending millions of dollars renovating their new Culver City headquarters, which they (sadly) renamed Sony Studios. But it could not have escaped their notice that because of a shortage of soundstages on that lot, they had been forced to rent studio space for two of their 1991 releases, *Bugsy* and *Hook*, just up Washington Boulevard—at the Culver Studios.

In 1991 Sony solved this problem by purchasing that studio as well. The purchase price was undisclosed but was estimated to be between $70 and $75 million (Tinker stated that it was, in fact, $80 million). The property was expected to eventually be used as a headquarters for Sony's television division, although Sony, like other multimedia corporations, probably quickly realized that the quicksilver nature of the entertainment industry would always make it hard to segregate their product by any singular delivery platform.

In 1995 out where 40 Acres once stood, and which is now 8620 Hayden Place, Los Angeles artist and historian Jim Heimann was commissioned by the city (Culver City) to create a piece called *King Kong Gone*. As the Culver City Public Art webpage describes it:

> The sculpture resembles a gate resting on a post with the title "King Kong Gone" boldly incorporated within its frame. The obscurity of the title within the context of the artwork's location is intended to stop passersby and draw them in for a closer examination. Incorporated into the artwork is the following explanation: "Near this site in 1927 the set for Cecil B. DeMille's *King of Kings* was built, used in 1932 as the gates for *King Kong*, which later became the burning City of Atlanta for *Gone with the Wind*, which was destroyed December 10, 1938."[13]

Studio Pass, a 1996 public art project by Jim Heimann, was built along the wall of Ince Boulevard and playfully depicts various studio insignia.

In 1996, and through an update in 1998, Heimann would also create *Studio Pass*, again for the Culver City Public Art program. This piece was attached to the wall of the studio on Ince Boulevard and whimsically depicted the studio logos of Ince, DeMille, Pathé, RKO, Selznick, Desilu, and the Culver Studios in wrought iron.

Sony, which never got to see its logo added to that wall, kept such a low profile that most of the preexisting Culver Studios signage, on the gates and internally, was left largely intact. In fact, like most of the previous owners, Sony ended up functioning primarily as a landlord, renting out the stages to more recording artists like Britney Spears for music videos and rehearsal space, and to other majors and both independent and internal productions, oftentimes only using the studio as an overflow plant for the main lot. Ultimately, Sony learned a hard lesson about modern-day Hollywood during this period that the managers of the other studios had long known, namely that the factory has to make a profit as well as the films.

Sony might have acquired the lot to service Columbia productions, but those productions had to pay to rent those stages (and those adjacent offices, production departments, and screening rooms), just as outside tenants would. So, if one of those Columbia productions, or one of those Columbia producers, could find a better rate at a competing studio, there was no compelling reason, other than corporate loyalty (seldom a deciding factor in Hollywood dealings), to shoot at home—at either home. And because of the overhead accrued by Sony's ownership of both lots, and the fact that both of those lots ultimately had to compete against one another for every single booking, Sony found itself constantly trying to underbid itself.

Undoubtedly shaking their heads at Hollywood's byzantine accounting practices, the Japanese giant sold Culver Studios in April of 2004 for $125 million, although Sony would remain a major tenant at the studio. The buyer this time was a private investment group known as Studio City Los Angeles.

These investors included the later-to-become-notorious Lehman Brothers and Pacifica Ventures, which was created to oversee the property. "About 90 percent of all the stage space in the world is in Southern California," said Dana Arnold, Pacifica Ventures' chief executive. "At any given time, the industry runs about 85 percent utilization. With real estate costs, it's hard to see any new production facilities being built any time soon."[14]

The chairman of Pacifica Ventures, Hal Katersky, like all of his predecessors, made promises about the renovations and refinements he was planning for the studio. Katersky also proposed a complex of offices and retail venues at the corner of Ince and Washington, which would be connected to the lot by a pedestrian bridge. Like many of these ambitious projects proposed over the years, this one never happened either.

One interesting addition to the lot that did happen was, most surprisingly, a backlot. Culver Studios had not had any standing outdoor sets since the 40 Acres era, although for the TV series *Mad About You* (1992–99), a small New York street had been constructed behind Stages 5 and 6 when the production got tired of the expense and inconvenience of having to lease similar sets at other, backlot-equipped studios. This set was later used for episodes of *Married with Children* (TV 1986–97), *Ned and Stacey* (TV 1995–97), and *The L Word* (TV 2004–9). *Playboy* also photographed the centerfold of their April 2004 playmate, Courtney Rachel Culkin, on the location. Unfortunately, during the production of the game show *Deal or No Deal* (TV 2005–9), this set was dismantled so that that production's trailers would have a place to park convenient to the stage.

The *Mad About You* and *Cougar Town* backlot sets that rose, and fell, behind Stages 5 and 6.

In 2010, however, for the production of the TV series *Cougar Town* (2009–15), a much more elaborate set was built in roughly the same place. "Florida Street," which portrayed the fictional Gulfhaven, Florida, was a delightfully pastel creation with a restaurant, a movie theater, a surf shop, and other Floridian-style facades, some of which included interiors. Though designed to be redressed for future productions, this apparently never happened.

Less attractive was an ugly legal dispute in 2007 between the studio's current owners and Katersky and his partner, Dana I. Arnold, who accused the duo of mismanagement of Culver Studios funds. Katersky fired back that the lawsuit was a ploy to derail another studio complex he was developing in Albuquerque, New Mexico, which Culver Studios management saw as a potential rival. The case ended with

40 Acres as an industrial park. Although this aerial was taken in 1988, it looks about the same today.

THE MEDIA CAMPUS 221

40 Acres, 2018. Tara was about where the big rig is on the right.

Pacifica Ventures officially removing itself from the management of Culver Studios and was finally settled in 2008.

That company's remaining management, led by the less than reliable, and now bankrupt and disgraced, Lehman Brothers, soldiered on until 2014, when the property was sold, yet again, to yet another investment group, Hackman Capitol Partners, for a reported $85 million. Speculation at the time held that the studio would be sold off piecemeal or redeveloped into high-end housing. Although the zoning on the property would have made that scenario problematic, the sad fate of 40 Acres certainly had proven that anything was possible.

Fortunately, this time it didn't happen. The *Hollywood Reporter*, in its coverage of the sale, reported:

> Rumors that the new owners have plans to convert parts of the landmarked studio into condominiums and shutter its 15 [sic] soundstages are unfounded, says Hackman CEO Michael Hackman. "We plan to retain and celebrate all of the historic aspects of the property, including the beautiful lawn and mansion. It's such a recognizable landmark. The notion that we would convert the lot into condos is baseless."[15]

In October 2017 commerce giant Amazon announced it was leasing more than 280,000 square feet of real estate on the lot, including the Mansion building, from Hackman. Amazon had been aggressively moving into the production of original filmed content for several years, and the studio, it was felt, would be a good fit as a base for its entertainment-related divisions. In 2018, with Amazon on the lot, a master redevelopment plan was once again drawn up to reconfigure the property and to build new facilities there for the future. Netflix, another web-based original content producer, had recently leased much of the Sunset Bronson Studios in Hollywood, which had once been home to Warner Bros., so Amazon's occupancy of the property was seen as timely, symbolic, and even, just maybe, appropriate.

On the other end of the lot, the one-time site of 40 Acres is now an industrial park. For example, an architectural firm, the Cunningham Group, now stands on the site of Tara and Jerusalem and Stalag 13. The Tarzan Jungle is now a practice site for the Culver City Fire Department. Strangely, the triangular shape of the property has been largely preserved. The same borders and backgrounds and landscapes that a thousand cinematographers stared at through their viewfinders are still there today. All that is missing are the sets.

Yet in some of the warehouses that stand where Mayberry and Manderley and Ince's Indians and *King Kong* and the *King of Kings* were once shot, as is the way in

40 Acres, 2018. Mayberry was about center.

Los Angeles, television shows are still occasionally produced. Century Studios (now defunct), Smashbox Studios, and Nant Studios have routinely rented out the buildings there for the production of projects like *ALF* (TV 1986–90), *The Wonder Years* (TV 1988–93), *Hell's Kitchen* (TV 2005–), *The Comeback* (HBO 2005, 2014), and *Silicon Valley* (TV 2014). The story continues. Somewhat.

When actor Don Knotts, Deputy Barney Fife from *The Andy Griffith Show*, died in 2006, some of his costars, including Griffith and Ron Howard, appeared on *Larry King Live* (TV 1985–2010) to discuss their friend. During the episode, the following exchange took place:[16]

> *Griffith:* "Mayberry was in Culver City. It was in a place called 40 Acres. Remember it, Ronnie?"
>
> *Howard:* "Absolutely. As a matter of fact, I took a drive by there about three weeks ago. I've known for a while that it's not there anymore, but, it was behind what was the old Selznick Studios, and it was great. It was kind of a playground for me. There's a great picture, in fact, of Don and I out there playing catch in 40 Acres that I have, in between setups, when I got interested in baseball and Don would throw the ball around with me every once in a while. It was a lot of fun."

In the end, for some of us, 40 Acres—whatever its actual size, whoever its owners, whether it was actually *in* Hollywood or not—is like those long-ago baseball games out on the backlot. The place threw something at us, and, in the audience, we caught it.

And we never threw it back.

Notes

Chapter 1: The Man with the Megaphone

1. Ince, "Drama and the Screen."
2. Gilbert, "Jack Gilbert Writes His Own Story."
3. *Moving Picture World*, "Ince to Build Studio in Culver City."
4. *Moving Picture World*, "Ince Culver City Studios Rapidly Near Completion."
5. *Moving Picture World*, "Thomas H. Ince to Return to Culver City."
6. Taves, *Thomas Ince*.
7. Thomas, "Ince Studios a Distinctive Achievement."

Chapter 2: Moguls, Millionaires . . . and a Bandleader

1. *Los Angeles Times*, "DeMille Adds to Plant Size."
2. Shepherd, *King Kong, Tara and Mayberry*.
3. Jewell, *RKO Radio Pictures*.
4. Eyman, *Empire of Dreams*.
5. Jewell, *RKO Radio Pictures*.
6. Haver, *David O. Selznick's Hollywood*.
7. Scheuer, "Hollywood Develops 'Ranch' Complex."
8. Harvey, "The Mystery of the Overlooked Answering Machine Message."
9. *Los Angeles Times*, "Boy, 13, Wounded by Police on Old Movie Set, Near Death."
10. *Los Angeles Times*, "Trailer Plant Damaged in Epidemic of Blazes."
11. Ibid.
12. *Hollywood Reporter*, "Selznick Studio Goes on Block."
13. Hecht, *A Child of the Century*.
14. Thompson, *Showman*.

15. Lasky, *RKO*.
16. Jewell, *Slow Fade to Black*.
17. *Daily Variety*, "Real Estate Developers Dickering Desilu for 40 Acres on Culver Backlot."
18. Desi Arnaz commentary in "Desilu Studios, an aerial view—Westinghouse promotion (1958)," YouTube video, 3:01, posted by VideoFun on June 8, 2015, www.youtube.com/watch?v=yaQeMzlQ_Zg.
19. Kaufman, "Desilu's '63–'64 Tenants Assure Record Billings of $22,000,000."
20. Kaufman, "Desilu's 2-Years-Ahead Cleanup of RKO Debt; $20,000,000 Written."
21. Ibid.
22. Blackburn, interview with author.
23. Ziarko, interview with author.

Chapter 3: Front Lot: Sets, Settings, and Set Pieces

1. O'Neil, Isabel O., with Donna Privitt. "Historic Resources Inventory—Laird International Studios," 1987.
2. *Los Angeles Sunday Times*, "The Sightly New Thomas H. Ince Studios at Culver City."
3. Chase, "'Strangers Kiss': The Final Wrap on a Tragic Love Affair."
4. Bertram, interview with author.
5. Cartnal, "Fashion Hoopla Over GWTW: He Dressed Scarlett O'Hara."
6. Myrick, *White Columns in Hollywood*.
7. *Los Angeles Times*, "Large Film Stage Near Completed."
8. Ebert, *Roger Ebert's Video Companion*.
9. Solow and Justman, *Inside Star Trek*.
10. Crowther, "Screen: The Searchers Find Action."
11. Creber, interview with author.
12. The Oliver Stone quote can be found on *The Hand* audio commentary track.

Chapter 4: 40 Acres: Hollywood Is a Facade

1. *Stage*, "40 Acre World."
2. Ibid.
3. *Daily Variety*, "Fire Destroys Culver City Set."
4. Jackovich and Sennet, "The Children of John Wayne, Susan Hayward, and Dick Powell Fear that Fallout Killed Their Parents."
5. Haver, *David O. Selznick's Hollywood*.
6. Pyron, *Southern Daughter*.
7. Ibid.
8. Tommy Jones paper published at www.tomitronics.com.
9. Ibid
10. *New York Times*, "David O. Selznick, 63, Producer of 'Gone with the Wind' Dies."
11. Shavin, "Desi and Lucy Will Restore Tara."
12. *Los Angeles Times*, "Revived Tara Gets Million-Dollar Look."
13. Jones, interview with author.
14. Walter O'Connor letter courtesy of Nancy Knechtel.

15. Peter Bonner is quoted from his website at www.savingtara.com.
16. Bonner, interview with author.
17. Turner and Goldner, *Spawn of Skull Island*.
18. Kurtz, "Technical Advisor: The Making of *Gone with the Wind*."
19. Haver, *David O. Selznick's Hollywood*.
20. Ibid.
21. Kurtz, "Technical Advisor: The Making of *Gone with the Wind*."
22. Haver, *David O. Selznick's Hollywood*.
23. *Denton (MD) Journal*, "Static."
24. The Andy Griffith quote can be found on www.visitmayberry.com.
25. Kelly, *The Andy Griffith Show*.
26. Kurtz, "Technical Advisor: The Making of *Gone with the Wind*."
27. The Jack Hill and Quentin Tarantino quotes can be found on the *Switchblade Sisters* DVD audio commentary track.

Chapter 5: The Media Campus

1. Scott, "Hugh O'Brian Partner in 'New' Movie Studio."
2. *Los Angeles Times*, "Blaze Destroys 2 Stages at Former Desilu Studio."
3. *Los Angeles Times*, "Culver City Studios Fire Spectacular, but Not Damaging."
4. *Los Angeles Times*, "Culver City Film Studio Hit by Fire."
5. Riney, "Dust, Memories Fill Landmark."
6. Ibid.
7. Ibid.
8. *Los Angeles Times*, "Initial Plans for Movie Studio Backlot Approved."
9. Faris, "Planners Approve Business Park on Old Desilu Lot."
10. Faris, "Culver Moves to Establish Movie Museum."
11. Tinker and Rukeyser, *Tinker in Television*.
12. Burbank, "Gannett Makes Entry Into Films with Top Bid for Laird Studios."
13. The Culver City Public Art site can be found at www.culvercity.org.
14. Gentile, "Sony Sells Culver Studios, Home to 'Lassie,' Classics."
15. Abramian, "Culver Studios Won't Become Condos, Says New Owner."
16. Andy Griffith and Ron Howard commentary in "Don Knotts Tribute—Larry King Part 1" YouTube video, 19:26, posted by opieopie28 on July 21, 2014, www.youtube.com/watch?v=mYNpwCFwopg.

Bibliography

Books

Aleiss, Angela. *Making the White Man's Indian*. Westport, CT: Praeger Publishers, 2005.

Beaucamp, Cari. *Joseph P. Kennedy Presents His Hollywood Years*. New York: Alfred A. Knopf, 2009.

Behlmer, Rudy. *Memo from David O. Selznick*. New York: Modern Library, 2000.

Bingen, Steven, Stephen X. Sylvester, and Michael Troyan. *M-G-M: Hollywood's Greatest Backlot*. Solana Beach, CA: Santa Monica Press, 2011.

———. *Paramount: City of Dreams*. Guilford, CT: Taylor Trade Publishing, 2017.

Bonner, Peter. *The Official Guide to the Saving Tara Project*. Marietta, GA: FirstWorks Publishing Co., 2014.

Bridges, Herb. *The Filming of Gone with the Wind*. Macon, GA: Mercer University Press, 1984.

Brownlow, Kevin. *The Parade's Gone By*. New York: Alfred A. Knopf, 1968.

Cerra, Julie Lugo, and Marc Wannamaker. *Movie Studios of Culver City*. Charleston, SC: Arcadia Publishers, 2011.

DeMille, Cecil B. *The Autobiography of Cecil B. DeMille*. Upper Saddle River, NJ: Prentice Hall, 1959.

Dunne, Gregory. *Monster: Living off the Big Screen*. New York: Vantage Books, 1998.

Ebert, Roger. *Roger Ebert's Video Companion*. 1996 ed. Kansas City, MO: Andrews and McMeel, 1996.

Eyman, Scott. *Empire of Dreams: The Epic Life of Cecil B. DeMille*. New York: Simon and Schuster, 2010.

Flamini, Roland. *Scarlett, Rhett, and a Cast of Thousands*. New York: Collier Books, 1975.

Frankel, Glenn. *The Searchers: The Making of an American Legend*. New York: Bloomsbury, 2014.

Friedrich, Otto. *City of Nets*. New York: Harper & Row, 1986.

Gabler, Neil. *An Empire of Their Own: How the Jews Invented Hollywood*. New York: Doubleday, 1988.

Haver, Ronald. *David O. Selznick's Hollywood*. New York: Alfred A. Knopf, 1980.

Hecht Ben. *A Child of the Century*. New York: Simon & Schuster, 1954.

Jewell, Richard B. *RKO Radio Pictures: A Titan is Born*. Los Angeles: University of California Press, 2012.

———. *Slow Fade to Black: The Decline of RKO Radio Pictures*. Los Angeles: University of California Press, 2016.

Jewell, Richard B., with Vernon Harbin. *The RKO Story*. New York: Arlington House, 1982.

Kelly, Richard. *The Andy Griffith Show*. Winston-Salem, NC: John F. Blair, Publisher, 1981.

Lasky, Betty. *RKO: The Biggest Little Major of Them All*. Upper Saddle River, NJ: Prentice-Hall, 1984.

Medved, Harry, and Michael Medved. *The Hollywood Hall of Shame*. New York: Perigee Books, 1984.

Morton, Ray. *King Kong: The History of a Movie Icon*. Montclair, NJ: Applause Cinema and Theater Books, 2005.

Myrick, Susan. *White Columns in Hollywood: Reports from the GWTW Sets*. Edited with an introduction by Richard Harwell. Macon, GA: Mercer University Press, 1982.

Presley, Cecilia deMille, and Mark A. Vieira. *Cecil B. DeMille: The Art of the Hollywood Epic*. Philadelphia: Running Press, 2014.

Pyron, Darden Asbury. *Southern Daughter: The Life of Margaret Mitchell*. New York: Oxford University Press, 1991.

Ramirez, Juan Antonio. *Architecture for the Screen*. Jefferson, NC: McFarland & Company, 2004.

Sanders, Coyne Steven, and Tom Gilbert. *Desilu: The Story of Lucille Ball and Desi Arnaz*. New York: Quill, William Morrow, 1993.

Selznick, David O. *Memo from David O. Selznick*. Selected and edited by Rudy Behlmer. New York: Viking Press, 1972.

Shepherd, Jacob N., III. *King Kong, Tara and Mayberry: The Story of the 40 Acres Backlot*. Amazon Digital Services, 2015.

Solow, Herbert, and Robert Justman. *Inside Star Trek*. New York: Pocket Books, 1996.

Taves, Brian. *Thomas Ince: Hollywood's Independent Pioneer*. Lexington: University Press of Kentucky, 2012.

Thompson, David. *Showman: The Life of David O. Selznick*. New York: Alfred A. Knopf, 1992.

Tinker, Grant, and Bud Rukeyser. *Tinker in Television: From General Sarnoff to General Electric*. New York: Simon & Schuster, 1994.

Turner, George E., and Orville Goldner. *Spawn of Skull Island: The Making of King Kong*. Expanded and revised by Michael H. Price and Douglas Turner. Baltimore, MD: Luminary Press, 2002.

Periodicals

Abramian, Alexandria. "Culver Studios Won't Become Condos, Says New Owner." *Hollywood Reporter*, March 5, 2014.

Burbank, Jeff. "Gannett Makes Entry into Films with Top Bid for Laird Studios." *Los Angeles Times*, December 16, 1986.

Cartnal, Alan. "Fashion Hoopla Over GWTW: He Dressed Scarlett O'Hara." *Los Angeles Times*, August 22, 1976.

Chase, Donald. "'Strangers Kiss': The Final Wrap on a Tragic Love Affair." *Los Angeles Times*, December 12, 1982.

Citron, Alan. "Sony to Acquire Culver Studios for Columbia Entertainment." *Los Angeles Times*, May 31, 1991.

Crowther, Bosley. "Screen: The Searchers Find Action; Entertaining Western Opens at Criterion 'Animal World' Shown at Little Carnegie." *New York Times*, May 31, 1956.

Daily Variety. "Conoco Leases Part of Desilu-Culver, Metro for Drilling." March 3, 1959.

———. "Desilu Expanding Culver City Lot." May 6, 1960.

———. "Fire Destroys Culver City Set." December 31, 1975.

———. "Fire Destroys '40 Acre' Lot." June 11, 1946.

———. "Real Estate Developers Dickering Desilu for 40 Acres on Culver Backlot." December 18, 1961.

———. "Selznick Sets RKO Prod'n Deal." September 2, 1955.

———. "$65,000,000 Tied Up in 24 Picture Studios—9,973 Work Daily—on Coast." January 4, 1928.

———. "Studio Building Cut by Defense." October 29, 1941.

———. "World Investment in Films over 2 1/2 Billions; $1,000,000,000 B.O. Last Year." October 13, 1967.

Denton (MD) Journal. "Static." September 25, 1964.

Faris, Gerald. "Culver Moves to Establish Movie Museum." *Los Angeles Times*, April 7, 1977.

———. "Planners Approve Business Park on Old Desilu Lot." *Los Angeles Times*, May 20, 1975.

Gentile, Gary. "Sony Sells Culver Studios, Home to 'Lassie,' Classics." *Orlando Sentinel*, April 7, 2004.

Gilbert, John. "Jack Gilbert Writes His Own Story." *Photoplay*, June 1928.

Harvey, Steve. "The Mystery of the Overlooked Answering Machine Message." *Los Angeles Times*, June 1, 2007

Hollywood Reporter. "Culver Studios Unveils New 'Cougar Town' Set." November 18, 2010.

———. "Selznick Studio Goes on Block." April 7, 1949.

Ince, Thomas. "Drama and the Screen." *Photoplay*, September 1916.

Jackovich, Karen G., and Mark Sennet. "The Children of John Wayne, Susan Hayward, and Dick Powell Fear That Fallout Killed Their Parents." *People*, November 10, 1980.

Jessen, J. C. "At the Ince Plant." *Motion Picture News*, January-February 1919.

Kaufman, Dave. "Desilu's '63–'64 Tenants Assure Record Billings of $22,000,000." *Daily Variety*, April 17, 1963.

———. "Desilu's 2-Years-Ahead Cleanup of RKO Debt; $20,000,000 Written." *Variety*, May 23, 1962.

Kurtz, Wilbur G. "Technical Advisor: The Making of Gone With the Wind—The Hollywood Journals of Wilbur G. Kurtz." Edited by Richard Harwell. *Atlanta Historical Society Journal*, Summer 1978.

Life. "The Day the Dream Factories Woke Up." February, 27, 1970.

Los Angeles Sunday Times. "The Sightly New Thomas H. Ince Studios at Culver City." December 1, 1918.

Los Angeles Times. "Blaze Destroys 2 Stages at Former Desilu Studio." December 12, 1971.

———. "Boulevard Work Pushed." November 13, 1927.

———. "Boy, 13, Wounded by Police on Old Movie Set, Near Death." December 3, 1938.

———. "Burned Stage to Be Rebuilt." July 20, 1927.

———. "Change Approved for Studio Zoning." December 5, 1974.

———. "Culver City Film Studio Hit by Fire." January 23, 1975.

———. "Culver City Studios Fire Spectacular, but Not Damaging." October 23, 1974.

———. "Culver Studio Purchased by Canada Firm." March 19, 1969.

———. "DeMille Adds to Plant Size." March 23, 1926.

———. "Federal OK Limits Gulf & Western Desilu Acquisition." July 26, 1967.

———. "Film Blaze Becomes Real as Wind Enters Studio." April 24, 1942.

———. "Film Concern Leases Studio." February 13, 1937.

———. "Firm Buys 2 Culver City Movie Plants." April 4, 1969.

———. "Industrial Park Zoning Urged for Desilu Site." August 22, 1974.

———. "Initial Plans for Movie Studio Backlot Approved." May 1, 1975.

———. "Large Film Stage Near Completed." September 11, 1927.

———. "Revived Tara Gets Million-Dollar Look." March 19, 1984.

———. "Selznick Takes Pathé Studio and Location Site." February 21, 1937.

———. "Studio Fire Loss $10,000." April, 27, 1948.

———. "This Week in Review." July 19, 1959.

———. "Trailer Plant Damaged in Epidemic of Blazes." May 20, 1943.

Moving Picture World. "Ince Culver City Studios Rapidly Near Completion." December 7, 1918.

———. "Ince Installs Weather Bureau." February 15, 1919.

———. "Ince to Build Studio in Culver City." December 8, 1917.

———. "Thomas H. Ince to Return to Culver City." July 6, 1918.

Musso, Joseph. "40 Acres: A History of the RKO Backlot Films." *Burroughs Bulletin*, April 1993.

New York Times. "David O. Selznick, 63, Producer of 'Gone with the Wind,' Dies." June 23, 1965.

Ng, David. "Amazon Studios Moving to Culver Studios in Culver City." *Los Angeles Times*, October 20, 2017.

Peltz, James F. "Light at the End of Chapter 11: Bankruptcy." *Los Angeles Times*, June 30, 1992.

Riney, Tom. "Dust, Memories Fill Landmark." *Santa Monica Evening Outlook*, November 17, 1970.

Scheuer, Thomas K. "Hollywood Develops 'Ranch' Complex." *Los Angeles Times*, October 18, 1936.

Scott, Vernon. "Hugh O'Brian Partner in 'New' Movie Studio." *Oakland (CA) Tribune*, June 4, 1969.

Shavin, Norman. "Desi and Lucy Will Restore Tara." *Atlanta Journal and Constitution Magazine*, December 14, 1958.

Silver Sheet. "The White House of Silent Drama." July 1923.

Stage. "40 Acre World." July 1936.

Thomas, Clark W. "Ince Studios a Distinctive Achievement." *Motion Picture News*, October 21, 1922.
Verrier, Richard. "Culver Studio Sale Is Near, Sources Say." *Los Angeles Times*, September 22, 2004.
Vincent, Roger. "Culver Studios Owners to Build Office Complex." *Los Angeles Times*, March 5, 2014.
Wanamaker, Marc. "Historic Hollywood Movie Studios, Part 1." *American Cinematographer*, March 1976.
———. "Historic Hollywood Movie Studios, Part 2." *American Cinematographer*, April 1976.
———. "Historic Hollywood Movie Studios, Part 3." *American Cinematographer*, May 1976.

Websites

"The Andy Griffith Show—40 Acres Backlot Today." www.youtube.com/watch?v=YDg8Ui8aR40&t=60s. A video tour of 40 Acres today.
Bison Archives. www.bisonarchives.com. Marc Wanamaker's Hollywood collections.
Culver City Historical Society. www.culvercityhistoricalsociety.org.
Culver City Public Art. www.culvercity.org/enjoy/about-culver-city/public-art.
Culver Studios. http://theculverstudios.com. The Culver Studios official website.
"Desilu Studios, an aerial view—Westinghouse promotion (1958)," YouTube video, 3:01, posted by VideoFun on June 8, 2015, www.youtube.com/watch?v=yaQeMzlQ_Zg.
"Don Knotts Tribute—Larry King Part 1" YouTube video, 19:26, posted by opieopie28 on July 21, 2014, www.youtube.com/watch?v=mYNpwCFwopg.
Film & TV Studio Backlots and Ranches. http://groups.yahoo.com/group/StudioBacklots. Lively studio backlot discussion site with much 40 Acres data.
"40 Acres": The Lost Studio Backlot of Movie & Television Fame. www.retroweb.com/40acres.html. Kipp Teague's one-stop treasury regarding all things 40 Acres.
Mayberry Historical Society. www.mayberry.info/history. A good place to start for all things Mayberry.
Memory Alpha. http://memory-alpha.wikia.com/wiki/Portal:Main. Comprehensive *Star Trek* reference site.
Mount Airy, North Carolina. www.visitmayberry.com. The website for the real-life inspiration for Mayberry can be found, tellingly, by typing "visit Mayberry" into a search engine.
Saving Tara. https://savingtara.com. Peter Bonner's site dedicated to the restoration and repair of the Tara facade.
Seeing Stars in Hollywood. www.seeing-stars.com. Jon Primrose's well-researched Hollywood visitors' site contains much pertinent material on studios.
Silent Locations. silentlocations.wordpress.com. John Bengston's groundbreaking location site.
Tomitronics. http://tomitronics.com. Tommy H. Jones's definitive account of Tara and the physical inspirations and legacy of *Gone with the Wind*.

Interviews

Bertram, John, assorted.
Blackburn, Billy, March 2018.
Bonner, Peter, February 2018.
Creber, William J., February 2018.
Howard, Ron, July 2018.
Jones, Tommy, March 2018.
McGee, Rex, January 2018.
Musso, Joseph, assorted.
Shepherd, Jacob N., III, assorted.
Teague, Kipp, assorted.
Wanamaker, Marc, assorted.
Ziarko, Charles, March 2018.

Video (special features consulted or quoted directly)

Citizen Kane (1941), 70th Anniversary Ultimate Collector's Edition (Warner Home Video, 2011).
Gone with the Wind (1939), The Scarlett Edition (Warner Home Video, 2009).
The Hand (1981), Twisted Terror Collection (Warner Home Video, 2007).
King Kong (1933), Two-Disc Collector's Edition (Warner Home Video 2006).
The Searchers (1957), Ultimate Collector's Edition (Warner Home Video 2007).
Switchblade Sisters (1975), Collector's Edition (Miramax Home Entertainment 2000).

Other

O'Connor, Walter. Undated personal letter to Nancy Knechtel, circa 1980s.
O'Neil, Isabel O., with Donna Privitt. "Historic Resources Inventory—Laird International Studios," 1987.
Robinson, W. W. *Culver City: A calendar of events in which is included, also, the story of Palms and Playa Del Rey together with Rancho la Ballona and Rancho Rincon de los Bueyes.* Los Angeles: Title Guarantee and Trust Company, 1939.
Wanamaker, Marc. "Chronology and History of the Studio—The Culver City Studios," 1991.

Index

A

Action (television program), 47, 85
Action in Arabia, 55
Adventure in Baltimore, 84, 186, 197
The Adventures of Jim Bowie (television program), 124, 134
The Adventures of Ozzie and Harriet (television program), 134, 186
The Adventures of Superman (television program), 142, 186, 197
The Adventures of Tom Sawyer
   music, 51
   scenes shot on Stage 3, 57
   scenes shot on Stage 4, 62
   scenes shot on Stage 8, 83
   scenes shot on Stage 11, 88
   scenes shot on Stage 14, 96
Against All Odds, 82
Aggie Appleby, Maker of Men, 88
Air Force One, 64
Airline Film & TV Promotions, 65, *65*
Airplane, 106
Aladdin, 92
ALF (television program), 224
Allan Quatermain and the Lost City of Gold, 60
Alley. *See* Bombed Town and Reform School Set
All Grown Up (television program), 79
All Mine to Give, 187
All of Me, 56
All Quiet on the Western Front, 120–21, *121*, 126, 199
The Amazing Westerbergs (television program), 79
Amazon, 223
American Creed, 55
The American President, 85
Andrews, Julie, 200–201
Androcles and the Lion, 95
The Andy Griffith Show (television program)
   Andy Griffith and Ron Howard on 40 Acres, 224
   Barracks Set and, 193
   City Streets lot and, 175, *176*, *177*, *178*
   Country Home Set and, 168
   Gone with the Wind sets used, 33, 168
   Goober's Service Station Set, 119, *119*
   Mayberry sets destroyed, 207
   Mayberry sets fire, 207
   realness of Mayberry, 177
   Reform School Set seen in, 202
   Residential Street Set and, *194*, 195
   Tarzan Jungle Set and, 125, 166
The Animal Kingdom, 120
Anna Christie, 12
Ann Vickers, 88, 182, 201
Arab Village and Medieval Village, 30, 73, *125*, 125–32, *126*, *128*, *129*, *131*
Argosy Pictures, 64
Armageddon, 107
Armendáriz, Pedro, 130
Armored Car Robbery, 59
Arnaz, Desi, 35, *37*, 109, 146 *See also* Desilu Studios
Arness, James, 132
Arnold, Dana I., 219, 221–22
Around the World, 55, 128

Around the World in 80 Days, 125, 187
Arrested Development (television program), 79
Ashworth, Carolyn, 150
Associated Producers, 12
The A-Team (television program), 106
Ateas Filmproduktion, 154
Atlanta Journal and Constitution, 146
At Sword's Point, 102
Attack, 202

B

Bachelor Party, 92
Back to Bataan, 101
Back to School, 56
Bad Company, 57, 179
Ball, Lucille, 35, 39, 146, 203 *See also* Desilu Studios
Ballona Creek and area
    All Quiet on the Western Front, 120–21, *121*
    described, 1
    oil and, 37
    size of 40 Acres and, 17, 18
    See also Tarzan Jungle
Banjo, 55, 142
Barbara Frietchie, 13
Bar Buckaroos, 85
Barker, Lex, 124
Barracks and Dock Sets, *191*, 191–94, *192*, *193*
Baskin-Robbins commercial, 64
Batman (television program)
    Ince Gate and, 117
    scenes shot at Arab Village and Medieval
      Village, 131
    scenes shot at City Streets, 188
    scenes shot at Country Home, 168
    scenes shot at Farm and Barn, 120
    scenes shot on *Gone with the Wind* set, 197
    scenes shot on Railroad Station Set, 168
    scenes shot on Stage 2, 56
    scenes shot on Stage 15, 103
    scenes shot on Stage 16, 106
    scenes shot on Western Street, 135
Bat Masterson (television program), 135
Baywatch (television program), 56, 217
Beach Pajamas, 62
Beatty, Warren, 154
Beauty and the Beast (television program), 86
Bedlam, 101
Bedlam of Beards, 182
Bed of Roses, 57, 201
Beetlejuice, 66, 86
Before Dawn, 93
Behind the Rising Sun, 86, 128
The Bells of St. Mary's, 64, 101
Beowulf, 87
Bergman, Ingrid, 83, 108

Berlin Express, 102, 202
Bertram, John, 49, 69
Best Friends, 91
Bette (television program), 79
Between Brothers (television program), 79
Beverly-Culver Studios, 205
Bewitched, 107
Beyond Victory, 63, 179, 199
Big City Productions, 65–66
The Big Gamble, 179
The Big Sky, 95
The Big Valley (television program), 143
Bill and Ted's Bogus Journey, 107
Bird of Paradise, 31, 57, 122, 157
Birdwell, Russell, 67
Blackbeard, the Pirate, 103, 192
Blackburn, Billy, 38
The Black Room, 126
Blades of Glory, 107
Blake Edwards Productions, 67
Blind Date, 63
Blondes by Proxy, 127
Blood Beach, 59
Blood on the Moon, 102
The Blue Danube, 46
The Blue Veil, 84
Bluhdorn, Charles, 203
Bodyguard, 84
Body of Evidence, 63
Bombardier, 97
Bombed Town and Reform School Set, 197, *198*,
    199, 199–202, *200*, *201*
Bonanza (television program), 135
Bond, Harry, 151
Bonner, Peter, 150
Boorman, John, 1
Born to Be Bad, 102
Born to Kill, 84
Born to Love, 182, 199
Botten, Gaylon, 127
Bowie, David, 212
Bride by Mistake, 59
Bride for Sale, 98, 186
Bringing Up Baby, 58
Brooks, Mel, 65
The Buddy Holly Story, 98
Bugsy, 82, 217
Building 6, 77–78
Building B, 65
Building E, 51
Building H, 66–67
Building I, 66
Building J, 69, 69–70, *70*
Building L, 70–72, *71*, *72*
Building M, 71–74, *73*

Building O, *74*, 74–75, 76
Building P, 52, 78–79
Building S, 66, *66*
Bungalow H, 51
Bungalow R, 68
Bungalow S, 68
bungalows and offices
   described, 64, 66, 68
   tenants, 64–68, *65*, *66*, *67*
   See also specific Buildings
Bungalow T, *67*, 68
Bungalow U, 68
Bungalow V, *67*, 68
A Bunny's Tale, 60
Butler, Robert, 98

C

The Cable Guy, 85
The Californians (television program), 134
Camera Effects Building, 77
Camp Henderson, 118–21, *119*, *120*, *121*
Can't Hurry Love (television program), 92
Capra, Frank, 64
Captain EO, 86
Captains Courageous, 44
Careful, My Love (television program), 193
Carl M. Beck Building Company of Los Angeles, 210, 212
Carnival Boat, 111
The Carol Burnett Show (television program), 95, 143
Carrie, 82
Car Shed. *See* Railroad Station Set
Casserini, Achille, 7, 16, 17, 18
Cat People, 90
Cecil B. DeMille Studios
   established, 15–16
   independent producers, 16
   office in Mansion, 19, *19*
   westerns and, 132
Century Studios, 224
Chances Are, 104
Channing, Carol, 132
Chaplin, Charlie, 13
Cheaper by the Dozen, 95
Cher . . . Special (television program), 98
Chicago, 174–75
Child's Play, 56
China Sky, 86, 129
Christmas with the Kranks, 93
Christopher Strong, 93
Cinematic Finance Corporation, 12
Citizen Kane
   props, 75
   scenes shot at the Mansion, 45
   scenes shot on City Streets Set, 184

   scenes shot on Stage 3, 58
   scenes shot on Stage 7, 81
   scenes shot on Stage 8, 83
   scenes shot on Stage 11, 90
   scenes shot on Stage 14, 97
City Heat, 84
City of Angels, 63
City Slickers, 92
City Slickers II: The Legend of Curly's Gold, 104
City Streets Set
   described, *173*, 173–74, *174*
   movies shot on, 174, 179, *180–81*, *182–84*, *183*, *184*, *185*, *186–87*, *189–91*, *190*
   television programs shot on, 175, *176*, *177*, 177–78, *178*, 187–89, *188*
The Cliffwood Avenue Kids (television program), 98
Climax (television program), 186
Cobra, 82
Collins, Joan, 189
The Comeback, 224
Commissary, 71–74, *73*
The Common Law, 82, 179
The Company She Keeps, 84, 86
The Conqueror, 98, 129–30
Contact, 92
Continental Oil Company (Conoco), 36–37
Conwell, Hillard, 19
Cooper, Merian C., 64, 156–57, 159
Cornered, 101
Cosmopolitan Productions, 12
Cougar Town (television program), *220*, 221
The Count of Monte Cristo, 121, 201
Country Home Set, 168, *169*, 170
Cracking Up, 82
The Craft, 99
The Creature Wasn't Nice, 92
Creber, William J., 104
The Crime Doctor, 182
Criminal Court, 84
Crimson Tide, 107
Crossfire, 63
Crowther, Bosley, 103
Cukor, George, 26, 161
Culkin, Courtney Rachel, 219
Culver, Harry Hazel, *16*
   background, 2
   Ince and, 7–8, 11
   interest in movie production, 2–3
Culver, Lillian, *16*
Culver City
   attempt at movie museum in Mansion, 210
   Fire Department practice site, 223
   Historical Society, 215
   Ince studio, 6
   layout, 29

rezoning of 40 Acres, 208, 210
Culver Studios, *204*, 205–6, 210, 212, 217
cutting room, 50

D

Daily Variety, 37
Dance, Girl, Dance, 94
A Date with the Falcon, 55
Davies, Marion, 12, 13
Davis, Frank I., 151
Davis, Marvin, 214
Day, Richard, 104
Days of Glory, 97
Dead Men Don't Wear Plaid, 59
Deal or No Deal (television program), 79, 219
Death Becomes Her, 107
de Havilland, Olivia, 67
DeMille, Cecil B., *15, 16*
    bible-themed movies, 150
    Building L and, 70
    building of Stage 11, 87–88
    Cecil B. DeMille Studios established, 15–16
    City Streets Set and, 174–75
    description of 40 Acres under, *17*
    independent producers and, 16
    movie production and profits, 12
    office in Mansion, 19, *19*
    Paramount and, 15, 21
    Producers Distributing Corporation, 15
    Reform School Set and, 197
    Stages 7, 8, and 9, 79
    westerns and, 132
Desilu Studios
    40 Acres under, 35–39, *38, 116*
    Mansion, *35*
    profitability, 37
    sale, 203
    shooting list (1963), *151*
    staffing, 37, 38
    Tara and, 146–47
    water tower, 109, *110*
    westerns and, 132
The Devil and Daniel Webster, 90, 142, 184
The Devil and Miss Jones, 90
Devotion, 179
Dial M for Murder (television program), 84
Dick Tracy Meets Gruesome, 63
The Dick Van Dyke Show (television program), 134, 135
Dietrich, Marlene, 126
Dock and Barracks Sets, *191*, 191–94, *192, 193*
Don't Tell Mom the Babysitter's Dead, 82
Dracula: Dead and Loving It, 56
dressing rooms, 107–8, *108*
Duel in the Sun, 132, *141*, 156

The Dynamite Brothers, 189

E

the Eagles, 212
Eastwood, Clint, 132
Eco Caters Café, 73
Edwards, Blake, 66
Edward Small's Reliance Pictures, 64
Eight Iron Men, 202
8mm, 104
Emergency Call, 182
The Enchanted Cottage, 101
The Entity, 98
Escape from New York, 103
Escape Plan, 104
E.T. the Extra-Terrestrial, 60, *60*
Every Girl Should Be Married, 82
Extant (television program), 62
Extreme Prejudice, 86
Eyman, Scott, 21

F

Fairbanks, Douglas, Jr., 205
The Falcon in Danger, 101
The Falcon in Mexico, 128
The Falcon's Adventure, 84, 101
Fallen, 56
The Fallen Sparrow, 81
Family Affair (television program), 189
A Family for Joe (television program), 79
Fantasia, 54
Farewell to Arms, 34
The Fargo Kid, 85
Farm and Barn Set, 118–21, *119, 120, 121*
The Farmer's Daughter, 55
Fat Albert, 99
Father Murphy, 86
Father Takes a Wife, 55
FBO Pictures Corporation, 20
Fenton, Harold, 159
A Few Good Men, 92
50 First Dates, 104
The Fighting Eagle, 120, 126
Fighting Father Dunne, 59, 186
The Fighting Generation, 90
The Fighting Kentuckian, 143
Fighting Love, 126
A Fine Mess, 95
fires, 31, 33
    Arab Village, 127
    final burning of Railroad Station Set, 168
    Gone with the Wind, 159–61, *160, 162, 163*
    in 1970s, 206–7, *207*
    Stage 6, 87
First National, 12

First Time Out (television program), 95
The First Traveling Saleslady (television program), 132
First Yank into Tokyo, 59, 124, 129
Flamingo Gold, 57
Flight for Freedom, 101
Flubber, 56
Flying Leathernecks, 67, 86
Footlight Fever, 58
The Forbidden Woman, 126
Ford, John, 35, 64, 103, 129–30
Forever and a Day, 94
The Fortune, 91, 154, 200
40 Acres
    amount of land, 16–17, 18
    area surrounding, 115
    battle for business with RKO Ranch, 22, 23
    constant changes, 17–18, 32, 42, 204
    Culver City rezoning, 208, 210
    current address, 217
    under DeMille, 16, 17, 19, 19
    under Desilu Studios, 35–39, 38, 116
    destruction of, 209, 212
    under Ince, 8–9, 9, 10, 11–12, 12, 13
    as industrial park, 221, 222, 223–24
    name, 18–19
    original owners, 7
    under RKO-Pathé, 17, 18, 20, 20–21, 21, 24, 25–26, 30–31
    under Selznick, 24–25, 27, 28, 28–29, 29, 30, 30–31, 117
    soundproofing, 28
    See also Ballona Creek
For Your Consideration, 47, 93
Foster, Julian, 147
Foster, Norman, 67
The Four Deuces, 190
Four Star Productions, 65
Francis, Anne, 111
Frank's Place (television program), 95
Frantz, James, 30
Fred Levinson Productions, 65–66
Fright Night, 82

G
Gable, Clark, 139, 201
Galaxy Quest, 107
The Gambler Returns: The Luck of the Draw (television program), 85
Gambling House, 55, 186, 192
A Game of Death, 59
Gannett (newspaper publisher), 213, 215–17
The Garden of Allah, 26, 96, 126, 159
gates, 51, 51, 115, 117, 118
Geffen Films, 66
General Tire and Rubber, 35

Genius at Work, 94
ghosts
    Building H, 66
    Mansion, 45
    Stage 1, 49
    Stage 2, 52–53
The Ghost Ship, 192
Ghosts of Mississippi, 86
Gideon's Trumpet (television program), 91
Gigli, 107
Gilbert, John, 4
Gildersleeve's Ghost, 97
Gilmore Girls (television program), 170
A Girl, a Guy and a Gob, 63
Girl Crazy, 182
The Girl Most Likely, 103, 195
Gleaming the Cube, 98
Glyn, Elinor, 13
The Godless Girl, 20, 197
Godzilla, 61
Golan, Menahem, 190
Goldwyn, Samuel, 16
Goldwyn Pictures, 7
Gomer Pyle, U.S.M.C. (television program), 117, 119, 120, 187, 193
Gone with the Wind
    cost, 163
    costumes, 70–71, 72
    "40 Acres" line in, 18–19
    Mansion and, 44
    music, 50
    Railroad Station and, 159–61, 160, 162, 163, 163–64, 164–65, 165
    scenes shot on City Streets Set, 183
    scenes shot on Reform School and Bombed Town Sets, 201, 201
    scenes shot on Residential Street Set, 195, 196
    scenes shot on Stage 2, 54
    scenes shot on Stage 4, 62
    scenes shot on Stage 7, 81
    scenes shot on Stage 8, 83
    scenes shot on Stage 9, 85
    scenes shot on Stage 10, 111
    scenes shot on Stage 11, 88, 89
    scenes shot on Stage 12, 93–94
    scenes shot on Stage 14, 96
    sets for, 30, 30–31, 33, 114, 114
    stars' quarters, 66, 66
    writing of, 26
    See also Tara
Good Sam, 63
Government Girl, 97
Gower Street lot (Hollywood), 20, 24
Grauman's Chinese Theater, 11
The Greatest Show on Earth, 59

The Greatest Story Ever Told, 105, *151*
    Arab Village and, 73, 131
    Jerusalem sets, 150–52
    scenes shot on Alley, 201
    scenes shot on Stage 14, 98
    scenes shot on Stage 16, 104, 106
    Stevens's offices, 65
The Great Jasper, 93
The Great Rupert, 186
Green Acres (television program), 168
Greene, Howard Duke, 26
The Green Hornet (television program), 78, 188
The Gregory Hines Show (television program), 79
Griffith, Andy, *176, 178, 179*
Griffith, D. W., 5, 156
GTG Entertainment, *213,* 215–17, *216*
Guestward Ho! (television program), 91, 168, 193
The Guns of Will Sonnett (television
    program), 135
Guys Like Us (television program), 95

H
Hackman, Michael, 222–23
Hackman Capitol Partners, 222–23
Hail the Woman, 46
Haley Tat Productions, 65
The Half-Naked Truth, 182
Halperin, Edward R., 64
Halperin, Victor, 64
The Hand, 106
Harryhausen, Ray, 49
Hart, William S., *2,* 5
Harum Scarum, 131
Hayward, Susan, 130
Hearst, William Randolph, 12–13
Hecht, Ben, 26
Heimann, Jim, 217–18, *218, 224*
Hell's Kitchen (television program), 224
Hepburn, Katharine, 182
Here We Go Again, 58
Herman, Alfred, 157
High and Dizzy, 186
Higuera, Francisco, 7
Higuera, Secundino, 7
Hilchey, Eugene, 49, 206
Hill, Jack, 189–90
Hirahara, Naomi, 29
His Kind of Woman, 91
Hitchcock, Alfred, *54,* 68
Hitler's Children, 101
H.K.M. Productions, 65–66
Hogan's Heroes (television program), 131, 135, *152,*
    152–54, *153,* 187, 194, 208
Hold 'Em Jail, 62
Holiday Affair, 102

Holler, Philip W., 11
Hollywood Reporter, 33, 222
Homegirl Café, 73
Honeymoon, 46, 81
Honeymoon in Vegas, 92
Hook, 92, 217
Hooray for Love, 93
House II: The Second Story, 60
House of Sand and Fog, 107
House Party, 82
Hughes, Howard, 34–35, 130
Human Highway, 95
Hunt the Man Down, 63
Hush, 79

I
If You Knew Susie, 59
I'll Be Seeing You, 51, 90, 184
I Love Lucy (television program), 35
Ilsa: She Wolf of the SS, 154
Image Steam Productions, 65–66
I'm Still Alive, 94
Ince, Elinor, 15
Ince, Thomas, *2, 11*
    Associated Producers and, 12
    background, 3
    Building L and, 70
    Cinematic Finance Corporation and, 12
    Culver and, 7–8, 11
    death, 13
    ghost of, 45
    on making movies, 3
    movie studio business model and, 4
    Triangle Film Corporation and, 5, 7
    See also Mansion; Thomas H. Ince Studios
Inceville, 4, 5, 5–6, *6*
Intermezzo: A Love Story, 62
    Dock Set and, 191, *192*
    music, 51
    scenes shot in Arab Village, 127
    scenes shot on Stage 2, 54
    scenes shot on Stage 3, 58
    scenes shot on Stage 7, 81
    scenes shot on Stage 8, 83
    scenes shot on Stage 11, 89
    scenes shot on Stage 12, 94
    scenes shot on Stage 14, 96
I Remember Mama, 102
The Iron Major, 102, 166
Isle of the Dead, 101
Is My Face Red, 179
I Spy (television program), 187
It's a Wonderful Life, 64, 91, 94, 97, 101
I Walked with a Zombie, 97, 128

J

Jackson, Michael, 212
Jake Spanner: Private Eye (television program), 85
The Jayne Mansfield Story, 59
The Jerk, 91
Jerry McGuire, 85
Jet Pilot, 192, 202
Jewell, Richard B., 20
Jewison, Norman, 66
Joe Dirt, 104
Johnny Angel, 191
Jones, Jennifer, 108, 141, *141*
Jones, Tommy, 140, 148, 150
Journey into Fear, 97
Judd for the Defense (television program), 189
The Judge Steps Out, 59
Judgment Night, 60
Julius Sizzler, 179
June First, 179

K

Katersky, Hal, 219, 221–22
Keith-Albee-Orpheum theater chain, 20
Kennedy, Joseph P., 20, 67
Kenny & Dolly: A Christmas to Remember (television program), 95
Kern, Hal, 26
The Kidnapping of the President, 106
Kill Bill: Vol 1, 104
Kincaid, Jack, 212, 215
The Kindred, 86
King Kong
    Arab Village and, 127
    artifacts from, 75, 76
    Jerusalem Temple and, 157, *158–59*, 159
    scenes shot on City Streets Set, 182
    scenes shot on Stage 12, 93
    scenes shot on Stage 14, 96
    Selznick and Cooper and, 156–57
    Stages 11,12, and 14 and, *206*
    Tarzan Jungle and, 122
King Kong Gone sculpture (Heimann), 217, *224*
The King of Kings, 17, 19, 52, 150, 151, *155*, 155–56
Kings Point Corp., 210
Kitty Foyle, 96
Klune, Ray, 31, 161, 164
Knotts, Don, *176*
Kosloff, Theodore, 13
Krippendorf's Tribe, 61
Kristofferson, Kris, 190
Kurtz, Wilbur, 161, 183

L

Ladies Man, 63
Lady Boss, 47
The Ladykillers, 99
Lady Luck, 184
The Lady Takes a Chance, 101
Lady with a Past, 83
Laird, Joseph R., 67, 71, 113–14, 210, 212–13
Laird International Studios, 210, *211*
Lancer (television program), 135
Land of the Giants (television program), 33, 189, 197
Larry King Live (television program), 224
Lasky, Betty, 34
Lassie (television program), 55, 187, 208
The Last Days of Pompeii, 127
The Last Ship (television program), 57
Las Vegas (television program), 53, 57
Laughlin, Tom, 65
The Leatherneck, 120, 201
Leeds, Martin N., 205
Legally Blonde, 82
Lehman Brothers, 219, 221–22
Leigh, Vivian, 66, *66*, *72*, 89, 161, *162*, 163, 165, *201*
Lepke, 190
Lesser, Sol, 122, 159
Let 'em Have It, 182
Let's Make Music, 58
Levin, Boris, 200
Liberty Productions/Films, 64
Life . . . and Stuff (television program), 79
Life Stinks, 99
Life with Bonnie (television program), 79
Little Lord Fauntleroy, 26
Little Men, 142, 166, 182–83
Little Women, 93, 94–95
Live from Baghdad (television program), 79
Livingston, Margaret, 13
Location Roads, 170–71, *171*, *172*
The Locket, 63
Lock Up, 98
Logan, Jacqueline, 13
Lonely Wives, 62, 111
The Long Night, 86, 102, 186
Los Angeles Sunday Times, 43
Los Angeles Times, 16, 27, 30, 31, 33, 148, 207, 208, 210, 214–15
Lost (television program), 120
The Lot (television program), 47, 85
Love and Marriage (television program), 79
Lucky Partners, 63
The Lucy Show (television program), 203
The Lusty Men, 98
The L Word (television program), 87, 219

M

Macao, 103, 129, 186
Mad About You (television program), 72, 92, 219, 220

Mad City, 60
Made for Each Other, 58, 62, 94, 96, 183
Madonna, 212
Magic Town, 63
The Magnificent Ambersons, 58, 86, 90, 184, *184*
The Main Event, 88
makeup department, *108*, 108–9
Malcolm & Eddie (television program), 79
Malice, 79
Man of the People (television program), 107
Mansion, *20*, *41*
    under Amazon, 223
    backside of, *45*
    building of, 7, 11
    Culver City attempt at movie museum in, 210
    Culver City Historical Society, 215
    under DeMille, 19, *19*, *45*
    described, *8*, *9*, *10*, 11, *11*, 43–44
    under Desilu, 35
    under Ince, *43*, 43–44, 45
    under Paramount, 203, *205*, *205*
    pool area, 45–46, *46*
    under RKO-Pathé Studios, *21*
    under Selznick, *25*, 44, *45*
    staircase, *44*
Man with a Camera (television program), 187
The Man with Two Brains, 60
Marilyn & Bobby: Her Final Affair, 63
Marine Raiders, 84
Married with Children (television program), 219
Mary Tyler More Show (television program), 213
Mass Appeal, 56
The Master Gunfighter, 65
Masters of the Universe, 92
Matchstick Men, 107
Matilda, 95
The Matrix, 107
Maxie, 103
Mayberry R.F.D. (television program), 119, *119*, 168, 170, *176*, *177*, 177–78, *178*
Mayer, Louis B., 33–34
McCartney, Paul, 212
McDonough, J. R., 24
McGee, Rex, 117
Medieval Village. *See* Arab Village and Medieval Village
Mellor, William C., 152
The Mentalist (television program), 170
Menzies, William Cameron, 161
Mexican Spitfire Out West, 94
Mexican Spitfire's Blessed Event, 58
Meyer, Gabriel S., 11
MGM (Metro-Goldwyn-Mayer)
    Continental Oil Company and, 36–37
    in Inceville, 7

    Southern Mansion, 142
    studios, 115, 208
    Tarzan movies, 122
Mighty Joe Young
    animation, 49
    scenes shot on Stage 7, 82
    scenes shot on Stage 8, 84
    scenes shot on Stage 11, 91
    scenes shot on Stage 12, 94
    scenes shot on Stage 14, 97
    scenes shot on Stage 15, 102
    Western Street and, 135, *136*
Mike's Murder, 106
Milbank, Jeremiah, 15, 16, 19, 21
Mill, 69, 69–70, *70*
Milwaukee Building Company, 11
The Mind of the Married Man (television program), 95
The Miracle of the Bells, 102, 109, 166, *185*, 186, 192
The Misconceptions (television program), 79
Mission Impossible (television program), 131, 154, 194, 197
Mitchell, Margaret, 138
Mitchell, Stephen, 147
The Monkey's Paw, 127, 191
Monster-in-Law, 87
Monster Squad, 86
The Moon and Sixpence, 31
Moorehead, Agnes, 130
The Morning After, 104
Morning Becomes Electra, 192
Morning Glory, 57, 182
Morosco, Oliver, 64
The Most Dangerous Game, 57, 75, 93, 96, 122
Motel Hell, 56
Mourning Becomes Electra, 91
Moving Picture World, 8
Moviola: The Scarlett O'Hara War (television program), 47, 143
Mr. Deeds, 92
Mr. Lucky, 101
Mr. & Mrs. Smith, 97
Mr. Saturday Night, 95
Mr. Woodcock, 79
music, 50–51
music videos, 212, 218
Musso, Joseph, 170
My Favorite Wife, 63
My Forbidden Past, 55, 86
My Life with Caroline, 100
My Pal Wolf, 97
Myrick, Susan, 73
My Three Sons (television program), 134, 135, 197, 202
My Wildest Dreams (television program), 79

N

The Nanny (television program), 79
Nant Studios, 224
National General Corporation, 65
Native Americans, actors portraying, 1, 2
The Natural, 106
The Navy Comes Through, 97
Ned and Stacey (television program), 219
Nelson, Argyle, 37
Netflix, 223
Never a Dull Moment, 63
Newman, Alfred, 26
The New People (television program), 120
Nichols, Mike, 154
Nicholson, Jack, 154
A Night of Adventure, 81
The Night of Love, 125–26, 156
The Night of the Hunter, 186–87
Night Song, 59
Nixon, 104
No, No, Nanette, 96
Nocturne, 63
None but the Lonely Heart, 83
North, 60
Nothing Sacred, 46, 85
NTA (National Telefilm Associates), 65
nuclear waste, 129–30

O

O'Brian, Hugh, 205
O'Connor, Walter, 148–49, 212, 215
Officer O'Brien, 57, 179
Official Detective (television program), 65
oil wells, 36–37
"Old Orchard," 18
"Old Stage 5," 77
Oliver Morosco Films, 64
Olivier, Laurence, *162*
O'Neil, Isobel, 40
O'Neil, Thomas F., 35
One Man's Journey, 182
One Minute to Zero, 91, 129
Open House, 57
OSF Industries LTD, 205
O'Sullivan, Maureen, 122
Out of the Past, 102
Out on a Limb, 104
Over the Top (television program), 79

P

Pacifica Ventures, 219, 221–22
Pacific Heights, 92
Palmerstown, USA, 56
Panama Flo, 81
Parachute Battalion, 94
The Paradine Case, 109
Paramount and DeMille, 15, 21
Paramount Television, 203, 205, *205*
Parents Wanted, 179
Parsons, Louella, 13
Passport to Destiny, 90
Past of Mary Holmes, 182
The Past of Mary Holmes, 81
Pathé Studio, 15–16, 20
Patty Hearst, 84
Payback, 82
Peck, Gregory, *141*
Peck's Bad Boy, 62
Pee-wee's Playhouse (television program), 79
Peggy Sue Got Married, 82
Pendleton, Robert C., 130
The Penguin Pool Murder, 88
Pennies from Heaven, 91
Pepin, Leo, 189, 206, 208
Perfect, 106
Perfect Film and Chemical Company, 203, 205
Peter Pan, 92
Petticoat Larceny, 55
Phillipps, H. G., 16
Philly, 62
Planes, Trains & Automobiles, 107
Plantation Club, 73
Play Girl, 97
Plunkett, Walter, 71, 72
Plymouth (television program), 99
The Polar Express, 62
Pony Express (television program), 135
Portrait of Jennie
   scenes shot on City Streets Set, 186
   scenes shot on Stage 2, 55
   scenes shot on Stage 3, 59
   scenes shot on Stage 7, 82
   scenes shot on Stage 10, 111
   scenes shot on Stage 14, 97
Postproduction building, 49–51, *50*
Powell, Dick, 130
Prestige, 93, 127
The Prisoner of Zenda, 88
Privitt, Dana, 40
Prizzi's Honor, 56
Producers Distributing Corporation (PDC), 15
producer's studios, described, 37
Property Building, *74*, 74–75, *76*
El Pueblo de Nuestra Señora la Reina de los Ángeles, 7

R

Race Street, 55
Rachel and the Stranger, 97
The Racket, 91, 112

Radioactive Dreams, 56
Raging Bull, 59
The Raid, 202
Railroad Depot. *See* Railroad Station Set
Railroad Station Set
    cost, 157
    films shot at, *155*, 155–57, *157*–59, 159, 166, *167*
    fire at, 206
    Gone with the Wind, 159–61, *160*, *162*, *163*, 163–64, *164*–65, *165*
    reason for 40 Acres and, 155–56
    television programs shot at, 166, 168
Railway Station. *See* Railroad Station Set
Rainbow Productions, 64
Raising Miranda (television program), 79, 216
Rancho Rincón de los Bueyes, 7
Rand, Ayn, 156
Rango (television program), 135
The Real McCoys (television program)
    City Streets Set, 187
    Farm and Barn Set, *119*
    oil and, 37
    scenes shot at Barracks Set, 193
    scenes shot at Jungle, 125
    scenes shot at Residential Street Set, 197
    scenes shot on Stage 7, 82
    scenes shot on Stage 8, 84
    storyline, 118–19
Rebecca
    Manderley Set, 139
    scenes shot in Arab Village, 127
    scenes shot on City Streets Set, 182
    scenes shot on Stage 1, 40, 48
    scenes shot on Stage 2, 54
    scenes shot on Stage 3, 58
    scenes shot on Stage 4, 62
    scenes shot on Stage 9, 85
    scenes shot on Stage 11, 89
    scenes shot on Stage 12, 94
    scenes shot on Stage 14, 96
    scenes shot on Western Street, 136
Rebound, 62, 111
Red Corner, 87
Reform School Set and Bombed Town, 197, *198*, *199*, 199–202, *200*, *201*
Regular Joe (television program), 79
Reiner, Rob, 66
rental studios, described, 37
Repli-Kate, 82
Residential Street Set, *193*, 194–95, *196*, 197
Return of Chandu, 122, 126, 159
Return to Mayberry (television program), 178–79
Reynolds, Debbie, 210
Rich and Famous, 103
Ricki Lake Show (television program), 79

Ride Beyond Vengeance, 125, 187
Riding in Cars with Boys, 99
Riff-Raff, 84
Right of Way (television program), 92
The Right to Romance, 93, 182
Ripley's Believe It or Not! (television program), 86
RKO-Pathé Studios
    40 Acres under, 17, *18*, 20, 20–21, *21*, 24, 25–26, 30–31
    Hughes and, 34–35
    Joseph Kennedy and, 20–21
    Property Building, 74
    Selznick lease of studio and backlot from, 24–25, 30–31
    Tarzan Jungle, 122, *123*, 124–25, 223
RKO Ranch (Encino), *22*, *23*, 24, 34, 132
The Roadhouse Murder, 182
Roar of the Dragon, 96, 127
Robertson-Cole Pictures, 5
Roberts Reality of the Bahamas LTD, 205
RoboCop, 66, 86
The Rock, 82
Rockabye, 182
The Rocket Man, 197
Rocky, 70
Roddenberry, Gene, 98
Rogers, Ginger, 132
Rolland D. Reed Productions, 65
Rookies in Burma, 90, 128
Roughshod, 102
Ruby Cairo, 82
Rush Hour 3, 107

S

The Saddle Buster, 81
The Saint in Palm Springs, 83
Samson and Delilah, 150
Santa Monica Evening Outlook, 208
Scarlett (television program), 146
Scene Dock building, *112*, 112–14, *113*, *114*
Schildkraut, Joseph, 152
Schoedsack, Ernest B., 159
Scott, Gordon, 124
The Searchers, 103
Secrets of the French Police, 85, 127
Selznick, Daniel, 210
Selznick, David O., *27*, *28*, 54
    assets for auction, 33, 70
    background, 25
    bungalows and, 64, *162*
    characteristics, 25, 27–28
    death, 34
    40 Acres under, 24–25, 27, 28, 28–29, *29*, 30, 30–31, *117*
    King Kong and, 156, 157

Mayer and, 33–34
music and, 50–51
office of, 26, *27*
set building and, 139–40
subletting of backlots, 31
Tara and, *140*, 141–42
United Artists and, 33
westerns and, 132
See also Gone with the Wind
Selznick, Irene, 26
Selznick, Lewis J., 25
Selznick, Myron, 25, 161
Selznick International Studio /Pictures, 25, 26, 33, 34
Sennett, Mack and Triangle Film Corporation, 5
sets, 4, 139–40 .*See also* specific locations
The Set-Up, 186
Seven Days Leave, 63
Seven Keys to Baldpate, 63
Seven Miles from Alcatraz, 58
The Seventh Victim, 81
Sex Court (television program), 79
Shasta McNasty (television program), 95
Shatner, William, 189
Shavin, Norman, 146
She, 159
She Couldn't Say No, 197
Shepherd, Jacob N., III (Jake), 7, 16–17
Short Circuit, 60
short subjects
   shot on City Streets, 179
   shot on Reform School and Bombed Town Sets, 201
   shot on Stage 2, 53–54, 55
   shot on Stage 7, 80
   shot on Stage 9, 85
Silicon Valley (television program), 224
Silver Sheet, 43
Sinbad the Sailor, 75
Since You Went Away, 58, 90, 94
Sin Takes a Holiday, 54
Sister Kenny, 101, 184
The Sky's the Limit, 101
Sledge Hammer! (television program), 92
Small, Edward, 64
Snug in the Jug, 88, 182
The Son of Kong, 93, 124
Son of Sinbad, 130
Sony Studios, 75, *76*, 217–19
Sony (Columbia) Studios, 7
The Sophisticated Gents (television program), 95
So This Is Washington, 184
Sour Grapes, 60
Space Camp, 106
Spanish Compound. *See* Arab Village and Medieval Village

Spellbinder, 92
Spellbound, 40, 63, 90, 101
Spelling, Aaron, 214
Spielberg, Steven, 75
The Spiral Staircase, 97
Stacey, Eric G., 73
Stack, Robert, 188
Stage 1, 47, *47*, 47–49, *48*, *49*
Stage 2, *52*, 52–57, *53*, *57*, 61, 77, 213
Stage 3, *52*, *52*, *53*, 57–62, *61*, 77, *77*
Stage 4, *52*, *52*, *53*, *61*, 62–64
Stage 5, 77, *77*, *78*, *78*, 79–80, 214
Stage 6, *77*, 77–78, *78*, 79, 87, 214
Stage 7, 80–82, *81*
Stage 8, 80, *81*, 82–85
Stage 9, 80, *81*, 85–87
Stage 10, 110–11, *111*
Stage 11, 87, 87–93
Stage 12, 87, 87–88, 93–95
Stage 14, 87, 87–88, 96–99
Stage 15, 88, 99, 99–104, *100*
Stage 16, 88, 99, 99–100, *100*
Stage magazine, 120, 125
Stallone, Sylvester, 66
Star!, 200
Starflight: The Plane That Couldn't Land (television program), 98
A Star Is Born (1937)
   Mansion and, 47
   scenes shot on Stage 7, 81
   scenes shot on Stage 11, 88
   scenes shot on Stage 12, 93
   scenes shot on Stage 14, 96
Starship Troopers, 61
Star Trek (television program), 28, 98, 131, 187
Star Wars: Episode VI—Return of the Jedi, 127
State's Attorney, 182
Station West, 192
Stealin' Home, 182
Steiner, Max, 50
Step by Step (television program), 79
Stevens, George, 65, 104, 106, 150–52
Stone, Oliver, 106
The Story of G.I. Joe, 94, 202
Stranger on the Third Floor, 89
Strangers Kiss, 56
Stranger Than Fiction, 87
A Streetcar Named Desire (television program), 103
Strictly Dynamite, 54
Stuart Little, 87, 99
Studio City Los Angeles, 219
Studio Pass sculpture (Heimann), 218, 218i
Suicide Fleet, 46, 96
Sunny, 88
Sunset, 47, 85

Sunset Action (television program), 47
Superman and the Mole-Men, 186
Swanson, Gloria, 21, 67
S.W.A.T., 104
Sweepstakes, 85
Sweet Hearts Dance, 66
Switchblade Sisters, 189–90, 201
Sylbert, Richard, 154
Syncopation, 81

T

Talmadge, Betty, 147, 148, 150
Tango & Cash, 104
Tara, *137, 138, 139, 140, 143, 144–45, 149*
   as allegory, 136
   building of, 140–41
   copied for home, 142–43
   current locations of components, 149–50
   current occupant of land, 223
   dismantled and shipped to Georgia, 147
   inspiration for, 147
   in novel, 138
   plantation sets at other studios, 142–43, 146
   set used in other productions, *141, 142,* 150–54, *151, 152, 153*
Tarantino, Quentin, 190
Tarzan and the Amazons, 124, 128
Tarzan and the Huntress, 124, 129
Tarzan and the Leopard Woman, 124, 129
Tarzan and the Mermaids, 124
Tarzan and the She-Devil, 124, 129
Tarzan and the Slave Girl, 124
Tarzan Jungle, 122, *123,* 124–25, 223
Tarzan's Desert Mystery, 124, 128
Tarzan's Hidden Jungle, 124
Tarzan's Magic Fountain, 124, 129
Tarzan's Peril, 124
Tarzan's Savage Fury, 124
Tarzan the Ape Man, 122
Tarzan the Fearless, 122
Tarzan Triumphs, 122, 127
Teaching Mrs. Tingle, 57
Teague, Kipp, 120, 195
Temple of Jerusalem. *See* Railroad Station Set
The Ten Commandments, 150
Tender Comrade, 101
Terminal Velocity, 107
The Texan (television program), 130, 132, 142, 187
That Girl (television program), 188
That Thing You Do!, 111
They Live by Night, 84, 186, 192
They Won't Believe Me, 59
The Third Man, 109
This Is Alice (television program), 65
This Land is Mine, 63

Thomas H. Ince Studios
   established, 7
   40 Acres under, 8, *8–9, 9, 10,* 11–12, *12, 13*
   lab, *50*
   promotional emblem, *14*
   westerns and, 132
   See also Mansion
Thompson, David, 33–34
Those Lips, Those Eyes, 84
Three Amigos!, 47, 95
Three Godfathers, 102
The Three Musketeers, 121
The Tick (television program), 95
Tinker, Grant, 46, 52, 66, 78, 110, 213–17
The Tip-Off, 57, 179
Tobor Pictures, 66
Tom, Dick and Harry, 58
Tom Brown's School Days, 89
Topaze, 57
Town and Country (television program), 60
Town & Country, 60, 61
The Toy Wife, 142
Trail Street, 84
Train Shed. *See* Railroad Station Set
Triangle Film Corporation, 5, 7
Triangle Studios, 6, 7
Triple Justice, 58
Tripoli, 202
TriStar, 66
True Confessions, 84
Turbulence, 91
Turkish Delight, 126
The Tuttles of Tahiti, 97, 124
TV 101 (television program), 60, 216
TWA, 65, *65*
Twelve Oaks. *See* Tara
20th Century Fox, 34, 208
The Two Jakes, 56
Two Tickets to Broadway, 55

U

Ulroy, Richard, 110
Undercover Blues, 92
Under the Rainbow, 47, 91, 143, 146
United Artists and Selznick, 33
The Untouchables (television program)
   Barracks Set and, 193
   Gone with the Wind set used, 33, 142, 166
   Mansion pool and, 46
   Residential Street Set, 197
   scenes shot on City Streets Set, 187, *188*
   scenes shot on Stage 2, 56
   scenes shot on Stage 15, 103
   scenes shot on Western Street Set, 135
Up Close & Personal, 85

Uptight, 91
Urson, Frank, 175
U.S. Marshal (television program), 119, 134–35
USA Today, 216

V

Valli, Alida, 109
The Van Dyke Show (television program), 92, 216
Vanguard Films, 33
Variety, 16, 33, 37
The Velvet Touch, 55
Verboten!, 142, 187, 202
Vern Gillian Productions, 65–66
V.I. Warshawski, 107
Vigilante Force, 190–91, *191*
Viva Zapata!, 129
Volk, A. G., 88

W

Wag the Dog, 47, 60
Walk Softly, Stranger, 55, 186
The Walter Winchell File (television program), 187
The Waltons (television program), 170
Wanamaker, Marc, *118*
Wardrobe, 70–72, *71*, *72*
Warner Bros., 66, 170
Watching Ellie (television program), 92
water tower, 109–10, *110*
Wayne, John, 35, 103, 129–30
Wednesday's Child, 182
Weissmuller, Johnny, 122, 124
Weitz, David, 33
Welles, Orson, 67, *67*
Western Street, 132, *132*, 133–34, 134–36, *136*, 207
Westward Passage, 121, 191–92
Whalley, Joanne, 146

What a Blonde, 94
What a Time, 82
What Price Hollywood?, 47, 93
What Women Want, 104
Wheeler, Lyle, *30*, 66, 159
When Every Day Was the Fourth of July (television program), 84
Where Danger Lives, 59, 166, 186
The Whip Hand, 95
Whirlybirds (television program), 65
White Tower, 82
The White Tower, 102
White Zombie, 64
Wilson, Al, 11
Window on Main Street (television program), 187, 197
Winter People, 47
WIOU (television program), 79, 216
Wise, Robert, *200*
The Wizard of Oz, 75
A Woman Commands, 111, 127
Woman on the Beach, 91
A Woman's Secret, 82, 84
The Wonder Years (television program), 224
Won Ton Ton: The Dog Who Saved Hollywood, 47, 72
Working Girl (television program), 79

Y

Yancy Derringer (television program), 134, 142
Yorkin, Bud, 66
You'll Find Out, 94
Young Bride, 85, 182
The Young in Heart, 46, 51, 111, 127
Yours, Mine & Ours, 87

Z

Ziarko, Charles, 38

About the Author

Steven Bingen has been acclaimed by *Cinema Retro* magazine as "one of the foremost Hollywood historians," having long worked within the motion picture industry, both in production and as a writer and archivist. He held a staff position at Warner Bros. Corporate Archive for fifteen years, aiding in the preservation and management of the studio's legend and legacy. He is an author of *MGM: Hollywood's Greatest Backlot*, *Warner Bros.: Hollywood's Ultimate Backlot*, *Paramount: City of Dreams*, and *The Stuff That Dreams Are Made Of: A Historical Look Behind the Gates of Warner Bros. Studios* and has contributed to many books and documentaries. He lives in Los Angeles, California.